NASA AERONAUTICS RESEARCH—AN ASSESSMENT

Committee for the Assessment of NASA's Aeronautics Research Program

Aeronautics and Space Engineering Board

Division on Engineering and Physical Sciences

NATIONAL RESEARCH COUNCIL
OF THE NATIONAL ACADEMIES

The National Academies Press
Washington, D.C.
www.nap.edu

THE NATIONAL ACADEMIES PRESS 500 Fifth Street, N.W. Washington, DC 20001

NOTICE: The project that is the subject of this report was approved by the Governing Board of the National Research Council, whose members are drawn from the councils of the National Academy of Sciences, the National Academy of Engineering, and the Institute of Medicine. The members of the committee responsible for the report were chosen for their special competences and with regard for appropriate balance.

This study was supported by Contract No. NASW-03009 between the National Academy of Sciences and the National Aeronautics and Space Administration. Any opinions, findings, conclusions, or recommendations expressed in this publication are those of the author(s) and do not necessarily reflect the views of the organizations or agencies that provided support for the project.

International Standard Book Number-13: 978-0-309-11913-9
International Standard Book Number-10: 0-309-11913-8

Available in limited supply from the Aeronautics and Space Engineering Board, 500 Fifth Street, N.W., Washington, DC 20001, (202) 334-2858.

Additional copies of this report are available from The National Academies Press, 500 Fifth Street, N.W., Lockbox 285, Washington, DC 20055; (800) 624-6242 or (202) 334-3313 (in the Washington metropolitan area); Internet, www.nap.edu.

Printed in the United States of America

THE NATIONAL ACADEMIES

Advisers to the Nation on Science, Engineering, and Medicine

The **National Academy of Sciences** is a private, nonprofit, self-perpetuating society of distinguished scholars engaged in scientific and engineering research, dedicated to the furtherance of science and technology and to their use for the general welfare. Upon the authority of the charter granted to it by the Congress in 1863, the Academy has a mandate that requires it to advise the federal government on scientific and technical matters. Dr. Ralph J. Cicerone is president of the National Academy of Sciences.

The **National Academy of Engineering** was established in 1964, under the charter of the National Academy of Sciences, as a parallel organization of outstanding engineers. It is autonomous in its administration and in the selection of its members, sharing with the National Academy of Sciences the responsibility for advising the federal government. The National Academy of Engineering also sponsors engineering programs aimed at meeting national needs, encourages education and research, and recognizes the superior achievements of engineers. Dr. Charles M. Vest is president of the National Academy of Engineering.

The **Institute of Medicine** was established in 1970 by the National Academy of Sciences to secure the services of eminent members of appropriate professions in the examination of policy matters pertaining to the health of the public. The Institute acts under the responsibility given to the National Academy of Sciences by its congressional charter to be an adviser to the federal government and, upon its own initiative, to identify issues of medical care, research, and education. Dr. Harvey V. Fineberg is president of the Institute of Medicine.

The **National Research Council** was organized by the National Academy of Sciences in 1916 to associate the broad community of science and technology with the Academy's purposes of furthering knowledge and advising the federal government. Functioning in accordance with general policies determined by the Academy, the Council has become the principal operating agency of both the National Academy of Sciences and the National Academy of Engineering in providing services to the government, the public, and the scientific and engineering communities. The Council is administered jointly by both Academies and the Institute of Medicine. Dr. Ralph J. Cicerone and Dr. Charles M. Vest are chair and vice chair, respectively, of the National Research Council.

www.national-academies.org

COMMITTEE FOR THE ASSESSMENT OF NASA'S AERONAUTICS RESEARCH PROGRAM

CARL J. MEADE, *Co-chair*, Northrop Grumman Integrated Systems, Santa Clarita, California
DONALD W. RICHARDSON, *Co-chair*, Donrich Research, Inc., West Palm Beach, Florida
RICHARD ABBOTT, Lockheed Martin Aeronautics Company, Palmdale, California
MEYER J. (MIKE) BENZAKEIN (NAE), Ohio State University, Columbus
JOHN T. (TOM) BEST, Arnold Engineering Development Center, Arnold Air Force Base, Tennessee
IAIN D. BOYD, University of Michigan, Ann Arbor
AMY L. BUHRIG, Boeing Commercial Airplanes, Renton, Washington
DAVID E. (ED) CROW (NAE), University of Connecticut, Glastonbury
FRANK L. FRISBIE, Apptis, Inc., Washington, D.C.
EPHRAHIM GARCIA, Cornell University, Ithaca, New York
PRABHAT HAJELA, Rensselaer Polytechnic Institute, Troy, New York
JOHN B. HAYHURST, The Boeing Company (retired), Kirkland, Washington
NANCY G. LEVESON (NAE),[1] Massachusetts Institute of Technology, Cambridge
ELI RESHOTKO (NAE), Case Western Reserve University (emeritus), Denver, Colorado
RAYMOND (RAY) VALEIKA, Delta Airlines (retired), Powder Springs, Georgia

Staff

ALAN ANGLEMAN, Study Director
SARAH CAPOTE, Program Associate

[1]Dr. Leveson resigned from the committee in May 2007.

AERONAUTICS AND SPACE ENGINEERING BOARD

Preface

The U.S. air transportation system is vital to the economic well-being and security of the United States. To support continued U.S. leadership in aviation, Congress and NASA requested that the National Research Council undertake a decadal survey of civil aeronautics research and technology (R&T) priorities that would help NASA fulfill its responsibility to preserve U.S. leadership in aeronautics technology. In 2006, the National Research Council published the *Decadal Survey of Civil Aeronautics*.[1] That report presented a set of six strategic objectives for the next decade of aeronautics R&T, and it described 51 high-priority R&T challenges—characterized by five common themes—for both NASA and non-NASA researchers.

The National Research Council produced the present report, which assesses NASA's Aeronautics Research Program, in response to the National Aeronautics and Space Administration Authorization Act of 2005 (Public Law 109-155). This report focuses on three sets of questions:

1. How well does NASA's research portfolio implement appropriate recommendations and address relevant high-priority research and technology challenges identified in the *Decadal Survey of Civil Aeronautics*? If gaps are found, what steps should be taken by the federal government to eliminate them?
2. How well does NASA's aeronautics research portfolio address the aeronautics research requirements of NASA, particularly for robotic and human space exploration? How well does NASA's aeronautics research portfolio address other federal government department/agency non-civil aeronautics research needs? If gaps are found, what steps should be taken by NASA and/or other parts of the federal government to eliminate them?
3. Will the nation have a skilled research workforce and research facilities commensurate with the requirements in (1) and (2) above? What critical improvements in workforce expertise and research facilities, if any, should NASA and the nation make to achieve the goals of NASA's research program?

[1]National Research Council. 2006. *Decadal Survey of Civil Aeronautics: Foundation for the Future*. Washington, D.C.: The National Academies Press. Available online at <http://www.nap.edu/catalog.php?record_id=11664>.

This report continues the good work begun by the *Decadal Survey of Civil Aeronautics*, and it expands that work to consider in more depth NASA aeronautics research issues related to the space program, non-civil applications, workforce, and facilities.

Carl Meade and Donald Richardson, *Co-chairs*
Committee for the Assessment of NASA's Aeronautics Research Program

Acknowledgment of Reviewers

This report has been reviewed in draft form by individuals chosen for their diverse perspectives and technical expertise, in accordance with procedures approved by the Report Review Committee of the National Research Council (NRC). The purpose of this independent review is to provide candid and critical comments that will assist the institution in making its published report as sound as possible and to ensure that the report meets institutional standards for objectivity, evidence, and responsiveness to the study charge. The review comments and draft manuscript remain confidential to protect the integrity of the deliberative process. We wish to thank the following individuals for their review of this report:

Graham Candler, University of Minnesota
Eric Feron, Georgia Institute of Technology
Awatef Hamed, University of Cincinnati
Pres Henne (NAE), Gulfstream Aerospace Corporation
Ilan Kroo (NAE), Stanford University
Andrew Lacher, MITRE Corporation
Lourdes Maurice, Federal Aviation Administration
Edmond Soliday, United Airlines (retired)
Dianne Wiley, The Boeing Company

Although the reviewers listed above have provided many constructive comments and suggestions, they were not asked to endorse the conclusions or recommendations, nor did they see the final draft of the report before its release. The review of this report was overseen by Martha Haynes, Cornell University, and Raymond S. Colladay, Lockheed Martin Astronautics (retired). Appointed by the NRC, they were responsible for making certain that an independent examination of this report was carried out in accordance with institutional procedures and that all review comments were carefully considered. Responsibility for the final content of this report rests entirely with the authoring committee and the institution.

Contents

APPENDIXES

Tables, Figures, and Box

TABLES

FIGURES

BOX

Summary

The United States is a leader in global aeronautics, and the National Aeronautics and Space Administration (NASA) has a critical role to play in preserving that position of leadership. NASA research facilities and expertise support research by other parts of the federal government and industry, and the results of research conducted and/or sponsored by NASA are embodied in key elements of the U.S. air transportation system, military aviation, and the space program. Maintaining a position of leadership in any field requires staying ahead of the competition by being the first to recognize and bridge each new gap into the future. This is generally a challenging task; were it not so, others would have overtaken the leader to set a faster pace. NASA aeronautics research can maintain a leadership position and carry on this tradition as long as its research is properly prioritized and research tasks are executed with enough depth and vigor to produce meaningful results in a timely fashion.

The National Research Council's (NRC's) *Decadal Survey of Civil Aeronautics: Foundation for the Future* (NRC, 2006) presents a set of six strategic objectives that the next decade of research and technology (R&T) should strive to achieve. It also describes the 51 highest-priority R&T challenges—characterized by five common themes—and an analysis of key barriers that must be overcome to reach the strategic objectives. Following the release of the *Decadal Survey of Civil Aeronautics*, the National Science and Technology Council (NSTC) released the *National Aeronautics Research and Development Policy* (NSTC, 2006). It then released the *National Plan for Aeronautics Research and Development and Related Infrastructure* a year later (NSTC, 2007). Although the *Decadal Survey of Civil Aeronautics* predated the National Policy and the National Plan, the strategic objectives defined in the *Decadal Survey* are closely aligned with the seven principles embodied in the NSTC documents (see Table S-1), and the ranking of the 51 highest-priority R&T challenges from the *Decadal Survey of Civil Aeronautics* remains valid.

NASA's aeronautics research is managed by the Aeronautics Research Mission Directorate (ARMD). The findings and recommendations in this report are based on a careful examination of NASA's research plans, the content of the *Decadal Survey of Civil Aeronautics,* the *National Aeronautics Research and Development Policy,* the *National Plan for Aeronautics Research and Development and Related Infrastructure,* and additional information regarding aeronautics research that NASA is or should be conducting to

TABLE S-1 Comparison of the Strategic Objectives from the *Decadal Survey of Civil Aeronautics* with the Principles from the *National Aeronautics Research and Development Policy* and the *National Plan for Aeronautics Research and Development and Related Infrastructure*

Strategic Objectives: *Decadal Survey*[a]	Principles: National Policy[b] and National Plan[c]
• Increase capacity.	• Mobility through the air is vital to economic stability, growth, and security as a nation.
•. Improve safety and reliability.	• Aviation safety is paramount.
• Increase efficiency and performance.	• Assuring energy availability and efficiency is central to the growth of the aeronautics enterprise.
• Reduce energy consumption and environmental impact.	• The environment must be protected while sustaining growth in air transportation.
• Take advantage of synergies with national and homeland security.	• Aviation is vital to national security and homeland defense. • Security of and within the aeronautics enterprise must be maintained.
• Support the space program.	
	• The United States should continue to possess, rely on, and develop its world-class aeronautics workforce.

[a]NRC (2006), p. 1.
[b]NSTC (2006), pp. 7-8.
[c]NSTC (2007), pp. 1-2.

support NASA space programs and other outside organizations, such the Federal Aviation Administration and the Department of Defense.

RESOURCES VERSUS SCOPE OF RESEARCH

NASA supports a great deal of worthwhile research. However, NASA must determine how to respond to a vast array of worthwhile research possibilities within the constraints of budget, facilities, workforce composition, and federal policies. The *Decadal Survey of Civil Aeronautics* (NRC, 2006) recommended that NASA use the 51 highest-priority R&T challenges in the *Decadal Survey* as the foundation for the future of NASA's civil aeronautics research program during the next decade. However, the *Decadal Survey* was designed to identify highest-priority R&T challenges without considering the cost or affordability of meeting the challenges.[1] As a result, even though the NASA aeronautics program has the technical ability to address each of the highest-priority R&T challenges from the *Decadal Survey* individually (through in-house research and/or partnerships with external research organizations), ARMD would require a substantial budget increase to address all of the challenges in a thorough and comprehensive manner.

In addition to resource limitations, NASA's aeronautics research program faces many other constraints (in terms of the existing set of NASA centers, limitations on the ability to transfer staff positions

[1]Other decadal surveys that the NRC routinely produces for NASA in the space sciences consider budgetary factors in formulating their findings and recommendations, and it may be worthwhile to follow that model in future decadal surveys for aeronautics research.

among centers, and limitations on the ability to compete with the private sector in terms of financial compensation in some critical fields), and attempting to address too many research objectives will severely limit the ability to develop new core competencies and unique capabilities that may be vital to the future of U.S. aeronautics.

Recommendation. The NASA Aeronautics Research Mission Directorate should ensure that its research program substantively advances the state of the art and makes a significant difference in a time frame of interest to users of the research results by (1) making a concerted effort to identify the potential users of ongoing research and how that research relates to those needs and (2) prioritizing potential research opportunities according to an accepted set of metrics. In addition, absent a substantial increase in funding and/or a substantial reduction in other constraints that NASA faces in conducting aeronautics research (such as facilities, workforce composition, and federal policies), NASA, in consultation with the aeronautics research community and others as appropriate, should redefine the scope and priorities within the aeronautics research program to be consistent with available resources and the priorities identified in (2) above (even if all 51 highest-priority R&T challenges from the *Decadal Survey of Civil Aeronautics* are not addressed simultaneously). This would improve the value of the research that the aeronautics program is able to perform, and it would make resources available to facilitate the development of new core competencies and unique capabilities that may be essential to the nation and to the NASA aeronautics program of the future.

USER CONNECTIONS

NASA civil aeronautics research will provide value to its stakeholders if and only if the results are ultimately transferred to industry, to the Federal Aviation Administration, and to the other organizations that manufacture, own, and operate key elements of the air transportation system. A closer connection between the managers of NASA aeronautics research projects and some potential users of NASA research would ensure that the need to transfer research results to users is properly considered in project planning and execution, and it would facilitate the formation of a coordinated set of research goals and milestones that are timed to meet the future needs of the nation. In addition, for technology intended to enhance the competitiveness of U.S. industry, U.S. leadership would be enhanced by a technology-transfer process that does not necessarily include the immediate, public dissemination of results to potential foreign competitors, so that the U.S. industrial base has a head start in absorbing the fruits of this research.

Recommendation. The NASA Aeronautics Research Mission Directorate should bridge the gap between research and application—and thereby increase the likelihood that this research will be of value to the intended users—as follows:

- Foster closer connections between NASA principal investigators and the potential external and internal users of their research, which include U.S. industry, the Federal Aviation Administration, the Department of Defense, academia, and the NASA space program.
- Improve research planning to ensure that the results are likely to be available in time to meet the future needs of the nation.
- Consistently articulate during the course of project planning and execution how research results are tied to capability improvements and how results will be transferred to users.
- For technology intended to enhance the competitiveness of U.S. industry, establish a more direct link between NASA and U.S. industry to provide for technology transfer in a way that

does not necessarily include the immediate, public dissemination of results to potential foreign competitors.

As part of the effort to implement this recommendation, NASA should ensure that the Next Generation Air Transportation System (NGATS/NextGen) Air Traffic Management (ATM)-Airportal Project and the NGATS ATM-Airspace Project meet the research and development (R&D) needs defined by the NextGen Joint Planning and Development Office (JPDO) for NASA.[2]

RESEARCH PLANNING AND ORGANIZATION

NASA's aeronautics research portfolio includes 10 projects, which are organized into three programs:

- Fundamental Aeronautics Program
  —Subsonic Fixed Wing (SFW) Project
  —Subsonic Rotary Wing (SRW) Project
  —Supersonics Project
  —Hypersonics Project
- Airspace Systems Program
  —NGATS ATM-Airportal Project
  —NGATS ATM-Airspace Project
- Aviation Safety Program
  —Integrated Vehicle Health Management (IVHM) Project
  —Integrated Intelligent Flight Deck (IIFD) Project
  —Integrated Resilient Aircraft Control (IRAC) Project
  —Aircraft Aging and Durability Project

In addition, ARMD manages the Aeronautics Test Program, which is intended to preserve key aeronautics testing capabilities.

NASA has developed a reference document for each of its 10 aeronautics research projects to define the rationale, scope, and detailed content of a comprehensive research effort to address each project area. NASA, however, does not consider these reference documents to be completed research plans, and in some cases they are difficult to correlate to the manner in which the projects are being implemented.

Recommendation. As reference documents and project plans are revised and updated, NASA should continue to improve the correlation between (1) the reference documents that describe project rationale and scope and (2) the project plans and actual implementation of each project.

MEETING THE CHALLENGES

The basic planning documents for most of NASA's research projects were prepared before the *Decadal Survey* was published in 2006, and the NASA research portfolio, as a whole, does not seem to have changed course in response to the *Decadal Survey*. Thus, the content of the *Decadal Survey of*

[2]The Next Generation Air Transportation System is now most commonly abbreviated as NextGen, but the titles of NASA's related research projects still feature the old acronym, NGATS.

TABLE S-2 Summary of How Well NASA's Aeronautics Research Supports the 51 Highest-Priority Research and Technology (R&T) Challenges from the *Decadal Survey of Civil Aeronautics*

Green = no significant shortcomings
Yellow = minor shortcomings
Black = major shortcomings
White = not relevant

ARMD --> Projects	Subsonic Fixed Wing	Subsonic Rotary Wing	Supersonics	Hypersonics	NGATS ATM-Airportal	NGATS ATM-Airspace	Integrated Vehicle Health Mgmt.	Integrated Intelligent Flight Deck	Integrated Resilient Aircraft Control	Aging Aircraft and Durability	Total Green	Total Yellow	Total Black	
ARMD --> Programs	*Fundamental Aeronautics Program*				*Airspace Sys. Prog.*		*Aviation Safety Program*				*Grade Summary by Challenge*			Titles of R&T Challenges (Some are abbreviated; see Table 1-1 for full titles.)
R&T Challenges in the Aerodynamics and Aeroacoustics Area														
A1											1	1	1	A1. Novel propulsion-airframe integration
A2											2	1	1	A2. Transition, boundary layer, and separation control
A3											1		2	A3. High performance and/or flexible multi-mission aircraft
A4a											3			A4a. Reduce aircraft and rotor noise
A4b											2	2		A4b. Prediction of performance of complex 3D configurations
A6	[a]										1	1		A6. Aerodynamics robust to atmospheric disturbances
A7a													1	A7a. Leverage advantages of formation flying
A7b											1	1	2	A7b. Wake vortex prediction, detection, and mitigation
A9												1	1	A9. V/STOL and ESTOL, including adequate control power
A10												1		A10. Reducing/mitigating sonic boom (novel aircraft shaping)
A11											1	3		A11. Robust and efficient multidisciplinary design tools
R&T Challenges in the Propulsion and Power Area														
B1a											2	1		B1a. Quiet propulsion systems
B1b												2		B2. Ultraclean gas turbine combustors
B3												2	2	B3. Intelligent engines and mechanical power systems
B4												2	1	B4. Improved propulsion system fuel economy
B5												1	1	B5. Propulsion systems for short takeoff and vertical lift
B6a												2		B6a. Variable-cycle engines to expand the operating envelope
B6b													4	B6b. Integrated power and thermal management systems
B8													1	B8. Propulsion systems for supersonic flight
B9													4	B9. Advanced aircraft electric power systems
B10											1			B10. Combined-cycle hypersonic propulsion systems
R&T Challenges in the Materials and Structures Area														
C1											1	1	1	C1. Integrated vehicle health management
C2												1	2	C2. Adaptive materials and morphing structures
C3											2	2	1	C3. Multidisciplinary analysis, design, and optimization
C4											1	2	1	C4. Next-generation polymers and composites
C5											1	1	1	C5. Noise prediction and suppression
C6a											1	2	2	C6a. Innovative high-temperature metals and environmental coatings
C6b											2	1	1	C6b. Innovative load suppression, and vibration and stability control
C8												1		C8. Structural innovations for high-speed rotorcraft
C9												5		C9. High-temperature ceramics and coatings
C10												2	3	C10. Multifunctional materials
R&T Challenges in the Dynamics, Navigation, and Control, and Avionics Area														
D1												2	2	D1. Advanced guidance systems
D2											2	1	1	D2. Distributed decision making and flight path planning
D3												1	2	D3. Aerodynamics and vehicle dynamics via closed-loop flow control
D4											1	1		D4. Intelligent and adaptive flight control techniques
D5											2	1	1	D5. Fault tolerant and integrated vehicle health management systems
D6												1	2	D6. Improved onboard weather systems and tools
D7													4	D7. Advanced communication, navigation, and surveillance technology
D8											3		1	D8. Human-machine integration
D9											1			D9. Synthetic and enhanced vision systems
D10													3	D10. Safe operation of unmanned air vehicles in the national airspace
R&T Challenges in the Intelligent and Autonomous Systems, Operations and Decision Making, Human Integrated Systems, Networking and Communications Area														
E1											1	3		E1. Design and evaluate complex interactive systems
E2												2		E2. Separating, spacing, and sequencing aircraft
E3												2		E3. Roles of humans and automated systems for separation assurance
E4													2	E4. Sensors, etc. to predict and measure wake turbulence
E5											1	1		E5. Information sharing among human and machine agents
E6												1	1	E6. Vulnerability analysis in the design of the air transportation system
E7												2		E7. Adaptive ATM techniques to minimize the impact of weather
E8a											2			E8a. Transparent and collaborative decision support systems
E8b											1	1		E8b. Operational and maintenance data to assess safety
E8c											1		1	E8c. Human operators in effective task and attention management
Totals for All 51 R&T Challenges from the Decadal Survey														
Green	10	4	6	1	7	2	3	3	2	0	38			
Yellow	9	13	9	5	3	7	3	3	1	5		58		
Black	8	14	9	5	6	7	2	0	1	1			53	

[a] Work on R&T Challenge A6 related to subsonic fixed wing aircraft is being done by the NASA Office of Safety.

Civil Aeronautics seems not to have been a significant factor in the selection of the research portfolio being pursued by many of ARMD's research projects. In any case, as illustrated in Table S-2, NASA is doing a mixed job in responding to the 51 highest-priority R&T challenges in the *Decadal Survey of Civil Aeronautics*. A summary follows.

There are no significant shortcomings in NASA's efforts to address four R&T challenges:[3]

- A4a. Aerodynamic designs and flow-control schemes to reduce aircraft and rotor noise
- B10. Combined-cycle hypersonic propulsion systems with mode transition
- D9. Synthetic and enhanced vision systems
- E8a. Transparent and collaborative decision support systems

Eight R&T challenges were uniformly evaluated as demonstrating minor shortcomings that could be corrected within the context of existing project plans:

- A10. Reducing/mitigating sonic boom (novel aircraft shaping)
- B2. Ultraclean gas turbine combustors
- B6a. Variable-cycle engines to expand the operating envelope
- C8. Structural innovations for high-speed rotorcraft
- C9. High-temperature ceramics and coatings
- E2. Separating, spacing, and sequencing aircraft
- E3. Roles of humans and automated systems for separation assurance
- E7. Adaptive ATM techniques to minimize the impact of weather

The committee verified NASA's own assessment that NASA is not supporting four R&T challenges:

- A7a. Aerodynamic configurations to leverage advantages of formation flying
- B9. High-reliability, high-performance, and high-power-density aircraft electric power systems
- D7. Advanced communication, navigation, and surveillance technology
- D10. Safe operation of unmanned air vehicles in the national airspace

The committee also determined that NASA is not substantively addressing three additional R&T challenges:

- B6b. Integrated power and thermal management systems
- B8. Propulsion systems for supersonic flight
- E4. Affordable new sensors, system technologies, and procedures to improve the prediction and measurement of wake turbulence

For the 32 other R&T challenges, NASA is effectively addressing some areas, but not others, and the overall assessment of these challenges is best described as "mixed." As shown in Table S-2, the committee assigned the following color-coded grades: a total of 149 green, yellow, or black grades—25 percent green, 39 percent yellow, and 36 percent black. Green means that a given project substantially

[3]The numbering of the challenges here and in Table S-2 is in accordance with the numbering scheme in the *Decadal Survey of Civil Aeronautics* (NRC, 2006).

meets relevant aspects of the intent of a particular R&T challenge and that the project will substantively advance the state of the art, with no significant shortcomings. Yellow means that a project has minor shortcomings in terms of its ability to support a given challenge, and those shortcomings are recoverable within the current overall project concept. Black means that a project has major shortcomings that would be difficult to recover from within the current overall project concept. White (or blank) means that the R&T challenge is not relevant to the project. The overall assessment for each challenge is indicated in the three columns labeled "Grade Summary by Challenge," which summarize the number of color-coded grades assigned to each challenge.

In a few cases, yellow or black grades indicate that NASA research plans are poorly conceived and that the resulting research will likely be ineffective. In most cases, however, yellow or black grades reflect inconsistencies between NASA project plans and the *Decadal Survey*. These inconsistencies are generally the result of NASA choosing to do little or no work in a particular task area and/or selecting research goals that fall short of advancing the state of the art far enough and with enough urgency either to make a substantial difference in meeting individual R&T challenges or the larger goal of achieving the strategic objectives of the *Decadal Survey of Civil Aeronautics*. However, as noted above, NASA does not have the resources necessary to address all 51 R&T challenges simultaneously in a thorough and comprehensive manner, and so it is inevitable that the project plans, as a whole, do not fully address all the priorities of the *Decadal Survey*.

NASA should respond to the shortcomings that are summarized in Table S-2 by implementing the recommendations in the preceding sections of this Summary.

WORKFORCE

There are—among NASA, the academic community, and the civilian aerospace industry—enough skilled research personnel to adequately support the current aeronautics research programs at NASA and nationwide, at least for the next decade or so. NASA may experience some localized problems at some centers, but the requisite intellectual capacity exists at other centers and/or in organizations outside NASA. Thus, NASA should be able to achieve its research goals, for example, by using NASA Research Announcements or other procurement mechanisms; through the use of higher, locally competitive salaries in selected disciplines at some centers; and/or by creating a virtual workforce that integrates staff from multiple centers with the skills necessary to address a particular research task. The content of the NASA aeronautics program, which has a large portfolio of tool development but little or no opportunities for flight tests, may in some cases hamper the ability to recruit new staff as compared with the space exploration program. In addition, there will likely be increased requirements for specialized or new skill sets. Workforce problems and inefficiencies can also arise from fluctuations in national aerospace engineering employment and from uneven funding in particular areas of endeavor.

Recommendation. To ensure that the NASA aeronautics program has and will continue to have an adequate supply of trained employees, the Aeronautics Research Mission Directorate should develop a vision describing the role of its research staff as well as a comprehensive, centralized strategic plan for workforce integration and implementation specific to ARMD. The plan should be based on an ARMD-wide survey of staffing *requirements* by skill level, coupled with an *availability* analysis of NASA civil servants available to support the NASA aeronautics program. The plan should identify specific gaps and the time frame in which they should be addressed. It should also define the role of NASA civil servant researchers vis-à-vis external researchers in terms of the following:

- Defining, achieving, and maintaining an appropriate balance between in-house research and external research (by academia and industry) in each project and task, recognizing that the appropriate balance will not be the same in all areas.
- Maintaining core competencies in areas consistent with (1) the highest-priority R&T challenges from the *Decadal Survey of Civil Aeronautics* and (2) NASA's role in the *National Aeronautics Research and Development Policy* and the *National Plan for Aeronautics Research and Development and Related Infrastructure*.
- Supporting the continuing education, training, and retention of necessary expertise in the NASA civil servant workforce and, as appropriate, determining how to encourage and support the education of the future aeronautics workforce in general.
- Developing, integrating, and applying foundational technology to meet NASA's internal requirements for aeronautics research.
- Defining and addressing issues related to research involving multidisciplinary capabilities and system design (i.e., research at Levels 3 and 4, respectively, as defined by ARMD).
- Ensuring that research projects continue to make progress when NASA works with outside organizations to obtain some of the requisite expertise (when that expertise is not resident in NASA's civil servant workforce).

NASA should use the National Research Council report *Building a Better NASA Workforce* (NRC, 2007) as a starting point in developing a comprehensive ARMD workforce plan.

FACILITIES

NASA has a unique set of aeronautics research facilities that provide key support to NASA, other federal departments and agencies, and industry. With very few exceptions, these facilities meet the relevant needs of existing aeronautics research. NASA also has a dedicated effort for sustaining large, key facilities and for shutting down low-priority facilities. However, some small facilities (particularly in the supersonic regime) are just as important and may warrant more support than they currently receive. In addition, at the current investment rate, widespread facility degradation will inevitably impact the ability of ARMD projects and other important national aeronautics research and development to achieve their goals.

Recommendation. Absent a substantial increase in facility maintenance and investment funds, NASA should reduce the impact of facility shortcomings by continuing to assess facilities and mothball or decommission facilities of lesser importance so that the most important facilities can be properly sustained.

REFERENCES

NRC (National Research Council). 2006. Decadal Survey of Civil Aeronautics: Foundation for the Future. Washington, D.C.: The National Academies Press. Available online at <http://www.nap.edu/catalog.php?record_id=11664>.

NRC. 2007. Building a Better NASA Workforce: Meeting the Workforce Needs for the National Vision for Space Exploration. Washington, D.C.: The National Academies Press. Available online at <www.nap.edu/catalog.php?record_id=11916>.

NSTC (National Science and Technology Council). 2006. National Aeronautics Research and Development Policy. Washington, D.C.: Office of Science and Technology Policy. Available online at <www.ostp.gov/html/NationalAeroR&DPolicy12-19-06.pdf>.

NSTC. 2007. National Plan for Aeronautics Research and Development and Related Infrastructure. Washington, D.C.: Office of Science and Technology Policy. Available online at <www.aeronautics.nasa.gov/releases/12_21_07_release.htm>.

1

Introduction

This report, which assesses aeronautics research conducted by the National Aeronautics and Space Administration (NASA), was prepared in response to the National Aeronautics and Space Administration Authorization Act of 2005 (Public Law 109-155), which directed NASA to enter into an arrangement with the National Research Council (NRC) for an assessment of aeronautics research. The specific purpose of this report is, in large part, to assess how well NASA's aeronautics research program is addressing the challenges and implementing the recommendations from the *Decadal Survey of Civil Aeronautics* (NRC, 2006). It is focused on answering three key questions from the statement of task:[1]

> 1. How well does NASA's research portfolio implement appropriate recommendations and address relevant highest-priority research and technology (R&T) challenges identified in the NRC *Decadal Survey of Civil Aeronautics*? If gaps are found, what steps should be taken by the federal government to eliminate them?
> 2. How well does NASA's aeronautics research portfolio address the aeronautics research requirements of NASA, particularly for robotic and human space exploration? How well does NASA's aeronautics research portfolio address other federal government department/agency non-civil aeronautics research needs? If gaps are found, what steps should be taken by NASA and/or other parts of the federal government to eliminate them? In order to answer this question the committee will identify and prioritize requirements for such research that fall within the scope of NASA's Aeronautics Research Program. To assist in the identification of such research requirements, NASA will provide the NRC with a list of its current research activities that contribute to these areas no later than March 12, 2007. It is likely that much of this research will be "dual use" or even "triple use," meaning that the research may provide benefit to the civil aeronautics community, and/or the space exploration community, and/or departments and agencies with non-civil aeronautics research needs.
> 3. Will the nation have a skilled research workforce and research facilities commensurate with the requirements in (1) and (2) above? What critical improvements in workforce expertise and research facilities, if any, should NASA and the nation make to achieve the goals of NASA's research program?

In answering the above questions, the committee that produced this report considered information contained in the *National Aeronautics Research and Development Policy* (NSTC, 2006), which was not available when the *Decadal Survey of Civil Aeronautics* was published. To a lesser extent, the

[1]The complete statement of task appears in Appendix A.

committee also considered information contained in the *National Plan for Aeronautics Research and Development and Related Infrastructure* (NSTC, 2007), which was released as this report was being finalized. As described below, this study considers how the principles contained in the National Policy and the National Plan might affect the ranking of R&T challenges in the *Decadal Survey*. However, it was beyond the scope of this study to validate the substance of the challenges contained in the *Decadal Survey* or to consider other R&T challenges not contained in that report, except in response to questions 2 and 3, above. Neither did this study attempt to assess the effectiveness of the management structure of the Aeronautics Research Mission Directorate (ARMD) or the current organization of ARMD research into various projects and programs, as described below.

OVERVIEW OF THE *DECADAL SURVEY OF CIVIL AERONAUTICS*

The *Decadal Survey of Civil Aeronautics* (NRC, 2006) presents a set of strategic objectives that the next decade of research and technology development should strive to achieve. It also provides a set of the highest-priority R&T challenges—characterized by five common themes—and an analysis of key barriers that must be overcome to reach the strategic objectives. The purpose of the *Decadal Survey* is to develop a foundation for the future—a decadal strategy for the federal government's involvement in civil aeronautics, with a particular emphasis on NASA's research portfolio.

The *Decadal Survey of Civil Aeronautics* also includes guidance on how federal resources allocated for aeronautics research should be distributed between in-house and external organizations, how aeronautics research can take advantage of advances in crosscutting technology funded by federal agencies and private industry, and how far along the development and technology readiness path federal agencies should advance key aeronautics technologies. It also provides a set of overall findings and recommendations to provide a cumulative, integrated view of civil aeronautics R&T challenges and priorities.

The *Decadal Survey* focuses on five areas that encompass the R&T of greatest relevance to civil aeronautics:

- Area A: Aerodynamics and aeroacoustics.
- Area B: Propulsion and power.
- Area C: Materials and structures.
- Area D: Dynamics, navigation, and control, and avionics.
- Area E: Intelligent and autonomous systems, operations and decision making, human integrated systems, and networking and communications.

The *Decadal Survey* then identifies and prioritizes within each area a set of key R&T challenges according to their ability to accomplish strategic objectives for U.S. aeronautics research. At the time the study was conducted, the federal government had yet to define what those strategic objectives should be. Therefore, in order to conduct the ranking, the authors of the *Decadal Survey* identified and defined six strategic objectives that, in their estimation, should motivate and guide the next decade of civil aeronautics research in the United States, pending the release of a national research and development (R&D) plan for aeronautics.[2] The six strategic objectives from the *Decadal Survey of Civil Aeronautics* are as follows (NRC, 2006, p. 1):

[2]In the same way, the research plans for the Next Generation Air Transportation System (NGATS) Air Traffic Management (ATM)-Airportal and ATM-Airspace Projects were prepared before the Next Generation Air Transportation System Joint Planning and Development Office (JPDO) had formally established R&D requirements. As a result the Airportal and Airspace Projects are a good-faith effort to meet expected JPDO requirements in both content and timing, pending release of an R&D

- Increase capacity.
- Improve safety and reliability.
- Increase efficiency and performance.
- Reduce energy consumption and environmental impact.
- Take advantage of synergies with national and homeland security.
- Support the space program.

A quality function deployment (QFD) process[3] was used to identify and rank-order a total of 89 R&T challenges in relation to their potential to achieve the above strategic objectives. The *Decadal Survey* recommends that NASA use the 51 highest-priority challenges as the foundation for the future of NASA's civil aeronautics research program during the next decade (see Table 1-1).

The *Decadal Survey of Civil Aeronautics* identifies several R&T challenges that are a high national priority, but they are not a high priority for NASA. This was the case if the challenge was poorly aligned with NASA's mission, if other organizations were likely to overcome the challenge, if NASA lacked the supporting infrastructure to investigate a particular challenge, and/or if the level of risk associated with the challenge was inappropriate for NASA research.[4] The following challenges from the *Decadal Survey* fall into this category (i.e., high national priority, but not a high NASA priority):[5]

- B11. Alternative fuels and additives for propulsion that could broaden fuel sources and/or lessen environmental impact
- B13. Improved propulsion system tolerance to weather, inlet distortion, wake ingestion, bird strike, and foreign object damage
- C11. Novel coatings
- C13. Advanced airframe alloys
- D11. Secure network-centric avionics architectures and systems to provide low-cost, efficient, fault-tolerant, onboard communications systems for data link and data transfer
- D13. More efficient certification processes for complex systems
- E11. Automated systems and dynamic strategies to facilitate allocation of airspace and airport resources
- E13. Feasibility of deploying an affordable broad-area, precision navigation capability compatible with international standards
- E17. Change management techniques applicable to the U.S air transportation system

Given the statement of task for this study, this report does not address NASA research as it relates to the above challenges or other challenges that are not included in Table 1-1 (except for four challenges that are addressed in Appendix C).

The *Decadal Survey* also makes eight recommendations (see Box 1-1) that summarize action necessary to properly prioritize civil aeronautics R&T and achieve the relevant strategic objectives.

---

requirements document by the JPDO. Likewise, the committee's assessments necessarily reflect the status of those projects at that point in their evolution.

[3]QFD is a group decision-making methodology often used in product design.

[4]The *Decadal Survey of Civil Aeronautics* assumes that risk is too low for NASA if it is so low that industry can easily complete the research, and the risk is too high if the scientific and technical hurdles are so high that there is very little chance of success.

[5]The numbering of the challenges here and throughout this report is in accordance with the numbering scheme in the *Decadal Survey of Civil Aeronautics* (NRC, 2006).

TABLE 1-1 Fifty-One Highest-Priority Research and Technology (R&T) Challenges for NASA Aeronautics, Prioritized by R&T Area

Area A. Aerodynamics and Aeroacoustics		Area B. Propulsion and Power		Area C. Materials and Structures		Area D. Dynamics, Navigation, and Control, and Avionics		Area E. Intelligent and Autonomous Systems, Operations and Decision Making, Human Integrated Systems, Networking and Communications	
A1	Integrated system performance through novel propulsion-airframe integration	B1a	Quiet propulsion systems	C1	Integrated vehicle health management	D1	Advanced guidance systems	E1	Methodologies, tools, and simulation and modeling capabilities to design and evaluate complex interactive systems
A2	Aerodynamic performance improvement through transition, boundary-layer, and separation control	B1b	Ultraclean gas turbine combustors to reduce gaseous and particulate emissions in all flight segments	C2	Adaptive materials and morphing structures	D2	Distributed decision making, decision making under uncertainty, and flight-path planning and prediction	E2	New concepts and methods of separating, spacing, and sequencing aircraft
A3	Novel aerodynamic configurations that enable high-performance and/or flexible multimission aircraft	B3	Intelligent engines and mechanical power systems capable of self-diagnosis and reconfiguration between shop visits	C3	Multidisciplinary analysis, design, and optimization	D3	Aerodynamics and vehicle dynamics via closed-loop flow control	E3	Appropriate roles of humans and automated systems for separation assurance, including the feasibility and merits of highly automated separation assurance systems
A4a	Aerodynamic designs and flow control schemes to reduce aircraft and rotor noise	B4	Improved propulsion system fuel economy	C4	Next-generation polymers and composites	D4	Intelligent and adaptive flight control techniques	E4	Affordable new sensors, system technologies, and procedures to improve the prediction and measurement of wake turbulence
A4b	Accuracy of prediction of aerodynamic performance of complex 3-D configurations, including improved boundary-layer transition and turbulence models and associated design tools	B5	Propulsion systems for short takeoff and vertical lift	C5	Noise prediction and suppression	D5	Fault-tolerant and integrated vehicle health management systems	E5	Interfaces that ensure effective information sharing and coordination among ground-based and airborne human and machine agents
A6	Aerodynamics robust to atmospheric disturbances and adverse weather conditions, including icing	B6a	Variable-cycle engines to expand the operating envelope	C6a	Innovative high-temperature metals and environmental coatings	D6	Improved onboard weather systems and tools	E6	Vulnerability analysis as an integral element in the architecture design and simulations of the air transportation system
A7a	Aerodynamic configurations to leverage advantages of formation flying	B6b	Integrated power and thermal management systems	C6b	Innovative load suppression, and vibration and aeromechanical stability control	D7	Advanced communication, navigation, and surveillance technology	E7	Adaptive ATM techniques to minimize the impact of weather by taking better advantage of improved probabilistic forecasts
A7b	Accuracy of wake vortex prediction, and vortex detection and mitigation techniques	B8	Propulsion systems for supersonic flight	C8	Structural innovations for high-speed rotorcraft	D8	Human–machine integration	E8a	Transparent and collaborative decision support systems
A9	Aerodynamic performance for V/STOL and ESTOL, including adequate control power	B9	High-reliability, high-performance, and high-power-density aircraft electric power systems	C9	High-temperature ceramics and coatings	D9	Synthetic and enhanced vision systems	E8b	Using operational and maintenance data to assess leading indicators of safety
A10	Techniques for reducing/ mitigating sonic boom through novel aircraft shaping	B10	Combined-cycle hypersonic propulsion systems with mode transition	C10	Multifunctional materials	D10	Safe operation of unmanned air vehicles in the national airspace	E8c	Interfaces and procedures that support human operators in effective task and attention management
A11	Robust and efficient multidisciplinary design tools								

NOTE: ATM, air traffic management; V/STOL, vertical and/or short takeoff and landing; ESTOL, extremely short takeoff and landing.
SOURCE: Reprinted from NRC (2006), Table ES-1.

BOX 1-1
Recommendations to Achieve Strategic Objectives for Civil Aeronautics Research and Technology, from the *Decadal Survey of Civil Aeronautics*

1. NASA should use the 51 challenges listed in Table ES-1 as the foundation for the future of NASA's civil aeronautics research program during the next decade.[a]
2. The U.S. government should place a high priority on establishing a *stable* aeronautics R&T plan, with the expectation that the plan will receive sustained funding for a decade or more, as necessary, for activities that are demonstrating satisfactory progress.
3. NASA should use five Common Themes to make the most efficient use of civil aeronautics R&T resources:

    - Physics-based analysis tools
    - Multidisciplinary design tools
    - Advanced configurations
    - Intelligent and adaptive systems
    - Complex interactive systems

4. NASA should support fundamental research to create the foundations for practical certification standards for new technologies.
5. The U.S. government should align organizational responsibilities as well as develop and implement techniques to improve change management for federal agencies and to assure a safe and cost-effective transition to the air transportation system of the future.
6. NASA should ensure that its civil aeronautics R&T plan features the substantive involvement of universities and industry, including a more balanced allocation of funding between in-house and external organizations than currently exists.
7. NASA should consult with non-NASA researchers to identify the most effective facilities and tools applicable to key aeronautics R&T projects and should facilitate collaborative research to ensure that each project has access to the most appropriate research capabilities, including test facilities; computational models and facilities; and intellectual capital, available from NASA, the Federal Aviation Administration, the Department of Defense, and other interested research organizations in government, industry, and academia.
8. The U.S. government should conduct a high-level review of organizational options for ensuring U.S. leadership in civil aeronautics.

[a]These 51 challenges are listed in Table 1-1 in the present report.

SOURCE: NRC (2006), p. 3.

The strategic objectives defined in the *Decadal Survey* are closely linked to the seven principles embodied in the *National Aeronautics Research and Development Policy* (NSTC, 2006) and the *National Plan for Aeronautics Research and Development and Related Infrastructure* (NSTC, 2007), which were released subsequent to publication of the *Decadal Survey*. Those principles are as follows (NSTC, 2006, pp. 7-8; NSTC, 2007, pp. 1-2):

- Aviation safety is paramount.
- Mobility through the air is vital to economic stability, growth, and security as a nation.
- Assuring energy availability and efficiency is central to the growth of the aeronautics enterprise.

TABLE 1-2 Comparison of the Strategic Objectives from the *Decadal Survey of Civil Aeronautics* with the Principles from the *National Aeronautics Research and Development Policy* and the *National Plan for Aeronautics Research and Development and Related Infrastructure*

Strategic Objectives: *Decadal Survey*[a]	Principles: National Policy[b] and National Plan[c]
• Increase capacity.	• Mobility through the air is vital to economic stability, growth, and security as a nation.
• Improve safety and reliability.	• Aviation safety is paramount.
• Increase efficiency and performance.	• Assuring energy availability and efficiency is central to the growth of the aeronautics enterprise.
• Reduce energy consumption and environmental impact.	• The environment must be protected while sustaining growth in air transportation.
• Take advantage of synergies with national and homeland security.	• Aviation is vital to national security and homeland defense. • Security of and within the aeronautics enterprise must be maintained.
• Support the space program.	
	• The United States should continue to possess, rely on, and develop its world-class aeronautics workforce.

[a]NRC (2006), p. 1.
[b]NSTC (2006), pp. 7-8.
[c]NSTC (2007), pp. 1-2.

- The environment must be protected while sustaining growth in air transportation.
- Aviation is vital to national security and homeland defense.
- Security of and within the aeronautics enterprise must be maintained.
- The United States should continue to possess, rely on, and develop its world-class aeronautics workforce.

As shown in Table 1-2 and detailed in Appendix C, there is good correlation between the strategic objectives of the *Decadal Survey* and the key principles of the National Plan; the only exceptions are "support for the space program" (which appears in the *Decadal Survey*) and "world-class aeronautics workforce" (which appears in the National Policy and the National Plan, although the National Plan includes no research goals or objectives related to this principle).[6] In addition, at the next level of detail, there are some differences between (1) the goals that the National Plan establishes to implement its principles and (2) the highest-priority R&T challenges that the *Decadal Survey* establishes for NASA to achieve the *Decadal Survey*'s strategic objectives. For example, the National Plan includes the following as a goal:

> Enable new aviation fuels derived from diverse and domestic resources to improve fuel supply security and price stability (NSTC, 2007, p. 27).

[6]The *National Plan for Aeronautics Research and Development and Related Infrastructure* notes that "aerospace workforce issues are being explored by the Aerospace Revitalization Task Force led by the Department of Labor pursuant to Public Law 109-420" (NSTC, 2007, p. 2).

The *Decadal Survey* notes the following:

> Challenge B11 (alternative fuels and additives for propulsion that could broaden fuel sources and/or lessen environmental impact) is clearly an important national priority. It was ranked lower as a NASA priority because the Department of Energy will need to take the lead in establishing the national infrastructure for an alternative fuel and because the combustion research needed to develop such a fuel will take much less time than putting an alternative fuel infrastructure in place. Furthermore, aviation fuels are likely to have a first call on petroleum supplies should they become scarce, so that the use of alternative fuels for aviation is likely to follow their widespread use for ground-based applications, which would place less stringent demands on weight, volume, reliability, safety, and certification of new systems and technologies. (NRC, 2006, p. 29)

Accordingly, even though the *Decadal Survey* ranked alternative fuels as a top-10 challenge in terms of national priority (in the Propulsion and Power area), alternative fuels failed to make the cut as a highest-priority challenge *for NASA*, and the *Decadal Survey* does not recommend that NASA take on alternative fuels as a high-priority research topic at this time.

This committee investigated whether the 51 highest-priority R&T challenges in the *Decadal Survey of Civil Aeronautics* remain valid in light of the National Policy and the National Plan. To do so, the 89 R&T challenges from the *Decadal Survey* were tentatively reranked on the basis of the principles in the National Policy and the National Plan. As detailed in Appendix C, only four R&T challenges not previously ranked among the top 51 moved into the top 51.

This does not mean that the list of the 51 highest-priority challenges should actually be adjusted, because the reranking described in Appendix C did not include additional deliberations on the merits and priorities of particular challenges, and those deliberations were a critical part of validating and finalizing the rankings in the *Decadal Survey*. Rather, the results of this exercise validate the rankings in the *Decadal Survey*, because it seems clear that even if the National Policy and the National Plan had been issued prior to the *Decadal Survey*, the rankings in the *Decadal Survey* would be essentially the same, with perhaps just a few changes for some of the R&T challenges that the *Decadal Survey* ranks near the dividing line between those challenges that were included in the group of the 51 highest-priority R&T challenges and those that were not.

Finding. The strategic objectives used to set research priorities in the *Decadal Survey of Civil Aeronautics*, the weighting of those objectives, and the ranking of research and technology challenges in the *Decadal Survey of Civil Aeronautics* are consistent with the R&D principles and priorities established by the *National Aeronautics Research and Development Policy* and the *National Plan for Aeronautics Research and Development and Related Infrastructure* that will implement the policy.

ORGANIZATION OF NASA'S AERONAUTICS RESEARCH

ARMD manages NASA's aeronautics research portfolio, which includes 10 projects organized into three programs:

- Fundamental Aeronautics Program
  —Subsonic Fixed Wing (SFW) Project
  —Subsonic Rotary Wing (SRW) Project
  —Supersonics Project
  —Hypersonics Project

- Airspace Systems Program
  - —Next Generation Air Transportation System (NGATS)[7] Air Traffic Management (ATM)-Airportal Project
  - —NGATS ATM-Airspace Project
- Aviation Safety Program
  - —Integrated Vehicle Health Management (IVHM) Project
  - —Integrated Intelligent Flight Deck (IIFD) Project
  - —Integrated Resilient Aircraft Control (IRAC) Project
  - —Aircraft Aging and Durability Project

In addition, ARMD manages the Aeronautics Test Program (ATP), which is intended to preserve key aeronautics testing capabilities.

Each of ARMD's 10 projects includes milestones at four levels:

Level 1. Foundational physics and modeling.
Level 2. Discipline-level capabilities.
Level 3. Multidisciplinary capabilities.
Level 4. System design.

The substance of each project is described in reference documents that NASA issued in May 2006 in support of a NASA Research Announcement (NRA) entitled "Research Opportunities in Aeronautics." Each document contains technical plans and milestones. The committee that produced this report reviewed the most current version of these documents (see NASA [2006a,b,c,d,e,f,g] and NASA [2007a,b,c]) and received briefings from the principal investigators (PIs) leading each of ARMD's 10 projects regarding the current status of the programs. The committee also familiarized itself with NRAs issued by the projects to solicit proposals for external research. However, the committee relied primarily on the reference documents and PI briefings to define the content of the projects, in part because the NRA solicitations are not a commitment to fund research in any particular area.

Finding. NASA's aeronautics program responds to many requirements related to many missions and in support of many users. For the most part, these requirements have not been formally established.

The *National Aeronautics Research and Development Policy* establishes high-level objectives, and the *National Plan for Aeronautics Research and Development and Related Infrastructure* takes this down to the next level of detail. The *Decadal Survey of Civil Aeronautics* includes a third level of detail, in the form of milestones associated with each R&T challenge, as well as additional recommendations. The committee supplemented its review of the above documents through discussions with other organizations—both within and outside NASA. These discussions led to a deeper understanding of what NASA's research requirements are (or should be) and an assessment of NASA's progress in meeting these requirements. Discussions were held with staff from the NASA Headquarters Aeronautics Research Mission Directorate, Exploration Systems Mission Directorate, Space Operations Mission Directorate, and Science Mission Directorate; each of NASA's 10 aeronautics research projects; congressional

---

[7]The Next Generation Air Transportation System is now most commonly abbreviated as NextGen, but the titles of NASA's related research projects still feature the old acronym, NGATS.

committees; the Office of Management and Budget; the Office of Science and Technology Policy; the Federal Aviation Administration (FAA); the Next Generation Air Transportation System (NextGen) Joint Planning and Development Office (JPDO); and the U.S. Air Force. The committee also relied on the expertise and experience of its members. For example, half of the members participated in authoring the NRC's *Decadal Survey* (to provide continuity with that study), and the other half were not veterans of the earlier effort (to ensure that the committee could view the issues and the *Decadal Survey* with a fresh perspective).

RESOURCE CONSIDERATIONS

Implementing all of the recommendations contained in the *Decadal Survey of Civil Aeronautics* and in the present report, *NASA Aeronautics Research: An Assessment,* will be very difficult, in part because, in accordance with the statement of task for the *Decadal Survey,* that earlier work identifies highest-priority R&T challenges without considering the cost or affordability of meeting the challenges. In addition, the committee that produced this new report requested from NASA comprehensive information on the NASA personnel and budget resources assigned to specific research tasks. Instead of providing this information, NASA directed the committee to assume that NASA would devote the resources necessary to accomplish the milestones described in the program plans that NASA provided to the study committee.

NASA must determine how to respond to a vast array of worthwhile research possibilities within the constraints of budget, facilities, workforce composition, and federal policies. The committee, while not constrained by these factors, was sensitive to them and attempted to craft its recommendations within reasonable bounds in all of these dimensions. Even so, the *Decadal Survey* and this assessment of NASA's aeronautics research would have been strengthened if the authoring committees had been directed to give some consideration to the likely cost and affordability of various challenges. This is normally the case when the NRC conducts decadal surveys for NASA related to space science. For example, cost realism was a "critical consideration" in developing the research strategy put forth in the last decadal on solar and space physics (NRC, 2003). The study that produced that report was designed "to ensure that its recommended research strategy is consistent with the anticipated budgets of the various federal agencies" (NRC, 2003, p. 53).

For whatever reasons, the *Decadal Survey of Civil Aeronautics* was designed such that cost realism was not included as a factor in setting its priorities. This report assesses whether NASA's plans for aeronautics research are consistent with the R&T challenges from the *Decadal Survey*, but it provides only a limited view of how well NASA is implementing those plans. For example, this committee obtained staffing levels for several projects, and the staffing levels assigned to the various elements of some projects seemed to be inconsistent with the proposed content of the research plan and the milestones contained therein. Other decadal surveys that the NRC routinely produces for NASA in the space sciences consider budgetary factors in formulating their findings and recommendations, and it may be worthwhile to follow that model in future decadal surveys for aeronautics research.

REPORT OVERVIEW

Chapter 1 of this assessment describes the context in which the report was written.

Chapter 2 evaluates how well each of NASA's 10 aeronautics research projects supports the 51 highest-priority R&T challenges from the *Decadal Survey of Civil Aeronautics,* as well as NASA's response to the eight overall recommendations in the *Decadal Survey*. In addition, the section entitled

"Space and Non-Civil Aeronautics Research" addresses NASA's own requirements for aeronautics research (including robotic and human space exploration) and the needs of other federal government departments and agencies for non-civil aeronautics research.

Chapter 3 addresses workforce and facility issues.

Chapter 4 identifies key gaps that must be eliminated for NASA's aeronautics program to meet key goals in terms of the R&T challenges from the *Decadal Survey of Civil Aeronautics* as well as internal NASA requirements for aeronautics research and the requirements that NASA is expected to satisfy in support of aeronautics research by other federal agencies. Chapter 4 also includes general recommendations for improving NASA's research.

Appendix A contains the study statement of task. Appendix B provides a short professional biography for each of the committee members. Appendix C evaluates how well each of NASA's 10 aeronautics research projects supports four R&T challenges that might have been included among the top 51 if the *National Aeronautics Research and Development Policy* and the *National Plan for Aeronautics Research and Development and Related Infrastructure* had been available in time to be considered in the preparation of the *Decadal Survey of Civil Aeronautics*. Appendix D is a list of acronyms.

In summary, implementing all of the recommendations contained in the *Decadal Survey of Civil Aeronautics* and in this report will be very difficult, because of constraints in terms of overall budget, the existing set of NASA centers, limitations on the ability to transfer staff positions among centers, and limitations on the ability to compete with the private sector in terms of financial compensation in some critical fields. Even so, NASA is already supporting a great deal of worthwhile research, and ongoing research will be very important, for example, to the critical work of the NextGen JPDO. The committee also recognizes that NASA's ARMD has staff who are talented, hardworking, and committed to excellence in aeronautics to serve the nation and the national aeronautics enterprise.

REFERENCES

NASA (National Aeronautics and Space Administration). 2006a. Aviation Safety Program, Aircraft Aging & Durability Project Technical Plan Summary. Washington, D.C.: NASA Headquarters, Aeronautics Research Mission Directorate. Available online at <www.aeronautics.nasa.gov/nra_pdf/aad_technical_plan_c1.pdf>.

NASA. 2006b. Aviation Safety Program, Integrated Intelligent Flight Deck Technical Plan Summary. Washington, D.C.: NASA Headquarters, Aeronautics Research Mission Directorate. Available online at <www.aeronautics.nasa.gov/nra_pdf/iifd_tech_plan_c1.pdf>.

NASA. 2006c. Fundamental Aeronautics, Hypersonics Project Reference Document. Washington, D.C.: NASA Headquarters, Aeronautics Research Mission Directorate. Available online at <www.aeronautics.nasa.gov/nra_pdf/hyp_proposal_c1.pdf>.

NASA. 2006d. Fundamental Aeronautics Program, Subsonic Fixed Wing Project Reference Document. Washington, D.C.: NASA Headquarters, Aeronautics Research Mission Directorate. Available online at <www.aeronautics.nasa.gov/nra_pdf/sfw_proposal_c1.pdf>.

NASA. 2006e. Fundamental Aeronautics, Subsonic Rotary Wing Reference Document. Washington, D.C.: NASA Headquarters, Aeronautics Research Mission Directorate. Available online at <www.aeronautics.nasa.gov/nra_pdf/srw_icp_response_c1.pdf>.

NASA. 2006f. Fundamental Aeronautics Program, Supersonics Project Reference Document. Washington, D.C.: NASA Headquarters, Aeronautics Research Mission Directorate. Available online at <www.aeronautics.nasa.gov/nra_pdf/sup_proposal_c1.pdf>.

NASA. 2006g. Next Generation Air Transportation System (NGATS) Air Traffic Management (ATM)-Airspace Project Reference Material, External Release Version. Version 6.0. Washington, D.C.: NASA Headquarters, Aeronautics Research Mission Directorate. Available online at <www.aeronautics.nasa.gov/nra_pdf/airspace_project_c1.pdf>.

NASA. 2007a. Aviation Safety Program, Integrated Resilient Aircraft Control: "Stability, Maneuverability, and Safe Landing in the Presence of Adverse Conditions." Washington, D.C.: NASA Headquarters, Aeronautics Research Mission Directorate. Available online at <www.aeronautics.nasa.gov/nra_pdf/irac_tech_plan_c1.pdf>.

NASA. 2007b. Aviation Safety Program, Integrated Vehicle Health Management Technical Plan Summary. Washington, D.C.: NASA Headquarters, Aeronautics Research Mission Directorate. Available online at <www.aeronautics.nasa.gov/nra_pdf/ivhm_tech_plan_c1.pdf>.

NASA. 2007c. Next Generation Air Transportation System (NGATS) Air Traffic Management (ATM)-Airportal Project Reference Material, External Release Version. Version 8.5. Washington, D.C.: NASA Headquarters, Aeronautics Research Mission Directorate. Available online at <www.aeronautics.nasa.gov/nra_pdf/airportal_project_c1.pdf>.

NRC (National Research Council). 2003. The Sun to the Earth—and Beyond: A Decadal Research Strategy in Solar and Space Physics. Washington, D.C.: The National Academies Press. Available online at <www.nap.edu/catalog.php?record_id=10477>.

NRC. 2006. Decadal Survey of Civil Aeronautics: Foundation for the Future. Washington, D.C.: The National Academies Press. Available online at <http://www.nap.edu/catalog.php?record_id=11664>.

NSTC (National Science and Technology Council). 2006. National Aeronautics Research and Development Policy. Washington, D.C.: Office of Science and Technology Policy. Available online at <www.ostp.gov/html/NationalAeroR&DPolicy12-19-06.pdf>.

NSTC. 2007. National Plan for Aeronautics Research and Development and Related Infrastructure. Washington, D.C.: Office of Science and Technology Policy. Available online at <www.aeronautics.nasa.gov/releases/12_21_07_release.htm>.

2

Challenges and Requirements for NASA Aeronautics Research

This chapter evaluates how well each of NASA's 10 aeronautics research projects supports the 51 highest-priority research and technology (R&T) challenges from the *Decadal Survey of Civil Aeronautics* (NRC, 2006) and NASA's own requirements for aeronautics research and the needs of other federal government departments and agencies for non-civil aeronautics research. The chapter also evaluates NASA's response to the eight overall recommendations that are contained in the *Decadal Survey.*

The evaluations of the 51 highest-priority R&T challenges are grouped according to the five areas from the *Decadal Survey*:

- Aerodynamics and Aeroacoustics
- Propulsion and Power
- Materials and Structures
- Dynamics, Navigation, and Control, and Avionics
- Intelligent and Autonomous Systems, Operations and Decision Making, Human Integrated Systems, and Networking and Communications

Appendixes A through E of the *Decadal Survey of Civil Aeronautics* contain lists of milestones for all of the challenges examined in the survey. The assessment of each R&T challenge below includes a list of the milestones established for that challenge. The purpose of this listing is to indicate the nature of the work that the *Decadal Survey* included within each challenge. However, the list of milestones for each challenge does not in all cases describe the complete scope of the challenge, as detailed in the *Decadal Survey of Civil Aeronautics.*

This chapter's evaluation of how well each of NASA's 10 aeronautics research projects supports the 51 highest-priority R&T challenges from the *Decadal Survey* is summarized in Tables 2-1 and 2-2. Each cell of Table 2-1 is color-coded with green, yellow, black, or white, as follows:

- *Green:* The project substantially meets relevant aspects of the intent of the R&T challenge and will substantively advance the state of the art, with no significant shortcomings.

TABLE 2-1 Summary of How Well NASA's Aeronautics Research Supports the 51 Highest-Priority Research and Technology (R&T) Challenges from the *Decadal Survey of Civil Aeronautics*

Green = no significant shortcomings
Yellow = minor shortcomings
Black = major shortcomings
White = not relevant

ARMD --> Projects	Subsonic Fixed Wing	Subsonic Rotary Wing	Supersonics	Hypersonics	NGATS ATM-Airportal	NGATS ATM-Airspace	Integrated Vehicle Health Mgmt.	Integrated Intelligent Flight Deck	Integrated Resilient Aircraft Control	Aging Aircraft and Durability	Total Green	Total Yellow	Total Black	
ARMD --> Programs	*Fundamental Aeronautics Program*				*Airspace Sys. Prog.*		*Aviation Safety Program*				*Grade Summary by Challenge*			Titles of R&T Challenges (Some are abbreviated; see Table 1-1 for full titles.)
R&T Challenges in the Aerodynamics and Aeroacoustics Area														
A1											1	1	1	A1. Novel propulsion-airframe integration
A2											2	1	1	A2. Transition, boundary layer, and separation control
A3											1		2	A3. High performance and/or flexible multi-mission aircraft
A4a											3			A4a. Reduce aircraft and rotor noise
A4b											2	2		A4b. Prediction of performance of complex 3D configurations
A6	[a]										1	1		A6. Aerodynamics robust to atmospheric disturbances
A7a													1	A7a. Leverage advantages of formation flying
A7b											1	1	2	A7b. Wake vortex prediction, detection, and mitigation
A9												1	1	A9. V/STOL and ESTOL, including adequate control power
A10												1		A10. Reducing/mitigating sonic boom (novel aircraft shaping)
A11											1	3		A11. Robust and efficient multidisciplinary design tools
R&T Challenges in the Propulsion and Power Area														
B1a											2	1		B1a. Quiet propulsion systems
B1b												2		B2. Ultraclean gas turbine combustors
B3												2	2	B3. Intelligent engines and mechanical power systems
B4												2	1	B4. Improved propulsion system fuel economy
B5												1	1	B5. Propulsion systems for short takeoff and vertical lift
B6a												2		B6a. Variable-cycle engines to expand the operating envelope
B6b													4	B6b. Integrated power and thermal management systems
B8													1	B8. Propulsion systems for supersonic flight
B9													4	B9. Advanced aircraft electric power systems
B10											1			B10. Combined-cycle hypersonic propulsion systems
R&T Challenges in the Materials and Structures Area														
C1											1	1	1	C1. Integrated vehicle health management
C2												1	2	C2. Adaptive materials and morphing structures
C3											2	2	1	C3. Multidisciplinary analysis, design, and optimization
C4											1	2	1	C4. Next-generation polymers and composites
C5											1	1	1	C5. Noise prediction and suppression
C6a											1	2	2	C6a. Innovative high-temperature metals and environmental coatings
C6b											2	1	1	C6b. Innovative load suppression, and vibration and stability control
C8												1		C8. Structural innovations for high-speed rotorcraft
C9												5		C9. High-temperature ceramics and coatings
C10												2	3	C10. Multifunctional materials
R&T Challenges in the Dynamics, Navigation, and Control, and Avionics Area														
D1												2	2	D1. Advanced guidance systems
D2											2	1	1	D2. Distributed decision making and flight path planning
D3												1	2	D3. Aerodynamics and vehicle dynamics via closed-loop flow control
D4											1	1		D4. Intelligent and adaptive flight control techniques
D5											2	1	1	D5. Fault tolerant and integrated vehicle health management systems
D6												1	2	D6. Improved onboard weather systems and tools
D7													4	D7. Advanced communication, navigation, and surveillance technology
D8											3		1	D8. Human-machine integration
D9											1			D9. Synthetic and enhanced vision systems
D10													3	D10. Safe operation of unmanned air vehicles in the national airspace
R&T Challenges in the Intelligent and Autonomous Systems, Operations and Decision Making, Human Integrated Systems, Networking and Communications Area														
E1											1	3		E1. Design and evaluate complex interactive systems
E2												2		E2. Separating, spacing, and sequencing aircraft
E3												2		E3. Roles of humans and automated systems for separation assurance
E4													2	E4. Sensors, etc. to predict and measure wake turbulence
E5											1	1		E5. Information sharing among human and machine agents
E6												1	1	E6. Vulnerability analysis in the design of the air transportation system
E7												2		E7. Adaptive ATM techniques to minimize the impact of weather
E8a											2			E8a. Transparent and collaborative decision support systems
E8b											1	1		E8b. Operational and maintenance data to assess safety
E8c											1		1	E8c. Human operators in effective task and attention management
Totals for All 51 R&T Challenges from the Decadal Survey														
Green	10	4	6	1	7	2	3	3	2	0	38			
Yellow	9	13	9	5	3	7	3	3	1	5		58		
Black	8	14	9	5	6	7	2	0	1	1			53	

[a] Work on R&T Challenge A6 related to subsonic fixed wing aircraft is being done by the NASA Office of Safety.

TABLE 2-2 Grade Summary for the 51 Highest-Priority R&T Challenges in the *Decadal Survey of Civil Aeronautics*, by Area

Report Area	Green	Yellow	Black	
Area A	12	11	8	Aerodynamics and Aeroacoustics
Area B	3	10	13	Propulsion and Power
Area C	8	18	12	Materials and Structures
Area D	9	7	16	Dynamics, Navigation, and Control, and Avionics
Area E	6	12	4	Intelligent and Autonomous Systems, Operations and Decision Making, Human Integrated Systems, and Networking and Communications
Total	38	58	53	

- *Yellow:* The project contains minor shortcomings, which are recoverable within the current overall project concept, such as the following:
  —Research described in the NASA task, if successful, would satisfy most, but not all, of the relevant aspects of the *Decadal Survey* R&T challenge (e.g., the task would make only moderate advances in the state of the art of relevant technologies, though the results would still be significant).
  —Research described in the NASA task, if successful, may not make a significant difference in a time frame of interest to users of the research results (e.g., because the level of effort is too low, or a different and more viable research approach should be selected, or some of the task is devoted to research goals inconsistent with the *Decadal Survey* R&T challenges, the aeronautics research requirements of NASA, and other federal government department or agency non-civil aeronautics research needs).
- *Black:* The project contains major shortcomings, which would be difficult to recover from within the current overall project concept, such as the following:
  —Research described in the NASA task, if successful, would make little or no progress in satisfying the *Decadal Survey* R&T challenge (e.g., the task would not make a significant advance in the state of the art of relevant technologies or the effort is meager compared to what is needed).
  —Research described in the NASA task, if successful, would be highly unlikely to make a significant difference in a time frame of interest to users of the research results (e.g., because the level of effort is too low, or a different and more viable research approach should be selected, or most of the task is devoted to research goals inconsistent with the *Decadal Survey* R&T challenges, the aeronautics research requirements of NASA, and other federal government department or agency non-civil aeronautics research needs).
  —The *Decadal Survey* R&T challenge is relevant to the NASA project, but the project is doing no related research.
- *White (or blank):* The R&T challenge is not relevant to the project.

As detailed in the discussion of individual challenges below, in a few cases yellow or black grades indicate that research plans developed by the Aeronautics Research Mission Directorate (ARMD) are poorly conceived and that the resulting research will likely be ineffective. In most cases, however, yellow or black grades reflect inconsistencies between NASA project plans and the *Decadal Survey*. These inconsistencies are generally the result of NASA choosing to do little or no work in a particular task area and/or selecting research goals that fall short of advancing the state of the art far enough and with enough urgency either to make a substantial difference in meeting individual R&T challenges or the larger goal of achieving the strategic objectives of the *Decadal Survey of Civil Aeronautics*. However, as noted in Chapter 4, NASA does not have the resources necessary to address all 51 R&T challenges simultaneously in a thorough and comprehensive manner, and so it is inevitable that the project plans, as a whole, do not live up to all of the expectations of the *Decadal Survey*.

The grades in Table 2-1 reflect the committee's assessment of how well a particular project addresses *relevant* aspects of a particular challenge. Thus, if only a small portion of a particular challenge is within the scope of a particular project but the project plans indicate that the project is or will do an excellent job in addressing that small research area, the cell in Table 2-1 representing the intersection of that project and challenge is green, even if the overall size of the relevant research is quite small. However, if a large portion of a particular challenge is within the scope of a particular project and if the project plans for the relevant research have minor or major shortcomings, the cell in Table 2-1 representing the intersection of that project and challenge is yellow or black, respectively, even if the overall size of the relevant research effort is quite substantial.

The difference between a black grade and a white grade is illustrated by R&T challenge A7a, Aerodynamic Configurations to Leverage Advantages of Formation Flying. None of ARMD's research projects plans to conduct research to support this R&T challenge, but if this challenge were pursued, the research would most appropriately be done by the Subsonic Fixed Wing (SFW) Project. Therefore, the SFW Project is graded black for R&T challenge A7a, and the other projects are graded white. As noted previously, NASA declined to provide detailed budget and staffing data for each project. Unless otherwise noted, the grades in Table 2-1 assume that the project research plans described to the committee will be implemented with enough funding and personnel resources to succeed. Thus, the grades primarily indicate the extent to which NASA's research plans are consistent with the *Decadal Survey of Civil Aeronautics,* but they do not necessarily indicate the likelihood that NASA will succeed in implementing those plans.

The overall assessment for each R&T challenge is indicated in the columns of Table 2-1 that summarize the number of grades assigned to each challenge, by color. As shown, the committee found no significant shortcomings in efforts by relevant ARMD research projects to address four R&T challenges (i.e., the grades assigned to these challenges are all green):

- A4a. Aerodynamic designs and flow-control schemes to reduce aircraft and rotor noise
- B10. Combined-cycle hypersonic propulsion systems with mode transition
- D9. Synthetic and enhanced vision systems
- E8a. Transparent and collaborative decision support systems

Eight R&T challenges received only yellow grades, indicating that ongoing work suffered from minor shortcomings that could be corrected within the context of existing project plans:

- A10. Reducing/mitigating sonic boom (novel aircraft shaping)
- B2. Ultraclean gas turbine combustors
- B6a. Variable-cycle engines to expand the operating envelope
- C8. Structural innovations for high-speed rotorcraft
- C9. High-temperature ceramics and coatings

- E2. Separating, spacing, and sequencing aircraft
- E3. Roles of humans and automated systems for separation assurance
- E7. Adaptive air traffic management (ATM) techniques to minimize the impact of weather

Seven R&T challenges received only black grades, indicating the presence of major shortcomings that would be difficult to recover from within the context of existing project concepts. The committee verified NASA's own assessment that NASA is not supporting four R&T challenges:

- A7a. Aerodynamic configurations to leverage advantages of formation flying
- B9. High-reliability, high-performance, and high-power-density aircraft electric power systems
- D7. Advanced communication, navigation, and surveillance technology
- D10. Safe operation of unmanned air vehicles in the national airspace

In addition, the committee has determined that NASA is not substantively addressing three other R&T challenges:

- B6b. Integrated power and thermal management systems
- B8. Propulsion systems for supersonic flight
- E4. Affordable new sensors, system technologies, and procedures to improve the prediction and measurement of wake turbulence

For the other 32 R&T challenges, as indicated in Table 2-1, NASA is effectively addressing some areas but not others, and the overall assessment of these challenges is best described as "mixed."

The grades for each row of Table 2-1 are explained in the sections that follow. In some cases, the comments for green grades are rather brief. Rather than prepare detailed assessments of areas where NASA is doing well (and significant corrective action is not required), the committee focused its attention on areas where improvements need to be made (those with black or yellow grades). Also, the committee chose not to justify its decision to assign white grades (that is, the determination that the scope of a given project was not relevant to a given R&T challenge).

AERODYNAMICS AND AEROACOUSTICS

This section summarizes the committee's assessment of NASA research related to the top 11 R&T challenges involving aerodynamics and aeroacoustics (Area A) in the *Decadal Survey of Civil Aeronautics* (NRC, 2006).

A1 Integrated system performance through novel propulsion-airframe integration

SFW	SRW	Supersonics	Hypersonics	Airportal	Airspace	IVHM	IIFD	IRAC	Aging A/C

This R&T challenge has the following milestones:

- Validate the predictive capability for three-dimensional (3-D) mean and dynamic distortion at the propulsion-airframe interface.
- Validate the predictive capability of the impact of reacting exhaust flows on external aerodynamics.

- Validate the predictive capability of acoustic radiation patterns from integrated propulsion-airframe configurations.
- Develop novel propulsion-airframe configurations for supersonic and hypersonic flight.

The SFW Project is investigating important innovative concepts related to this R&T challenge. Research goals include development of dynamic models of integrated control systems, development and application of prototype actuators and innovative control methods, and laboratory experiments and piloted simulations to validate closed-loop system performance. Plans include flight-test validation of predictive models for propulsion-airframe integration of unconventional vehicle configurations, such as the blended-wing-body (BWB) aircraft, where possible.

The Supersonics Project is supporting extensive code development in this area, and NASA plans to rely on yet-to-be-established partnerships with industry to execute key aspects of the above milestones with regard to validation. However, NASA has not established any notional vehicles to help refine its work. This shortcoming could be addressed, perhaps, by using one of the vehicle concepts developed by the Defense Advanced Research Projects Agency (DARPA) as part of the Quiet Supersonic Platform (Wlezien and Veitch, 2002), modified as necessary to reflect civil rather than military performance requirements.

This R&T challenge is focused on novel configurations for propulsion-airframe integration. Propulsion-airframe integration is a key component of any air-breathing hypersonic vehicle. In addition, the Vehicle Technology Integration, Propulsion Technology Integration, and Physics Based Multidisciplinary Design, Analysis, and Optimization (MDAO) elements of the Hypersonics Project are focused on the development of predictive tools. However, the Hypersonics Project will not validate the performance of these tools, nor will it exercise the tools to investigate the design of any specific, novel propulsion-airframe configurations.

A2 Aerodynamic performance improvement through transition, boundary-layer, and separation control

SFW	SRW	Supersonics	Hypersonics	Airportal	Airspace	IVHM	IIFD	IRAC	Aging A/C

This R&T challenge has the following milestones:

- Develop energy-efficient and flexible active flow-control actuators.
- Develop improved models for the operation of flow actuators.
- Demonstrate techniques to incorporate these models into flow-simulation schemes.
- Validate models and simulation schemes through comparison with experiments.

The SFW Project is investigating smart material actuators as well as active and passive control concepts. Research plans include validation at the configuration, component, and physics levels. Plans include a key test to demonstrate improved performance via a high-lift wind tunnel model with flow-control actuation integrated into realistic aircraft structure.

The SRW Project has no planned research to address the above milestones.

The Supersonics Project is working on both foundational research and performance improvement related to this challenge, including actuator development. This work would be facilitated if NASA had a quiet wind tunnel in the Mach number range being investigated (approximately Mach 1.5 up to Mach 2.5) for transition validation. Based on experience with existing quiet wind tunnels at Langley Research Center (which operates at Mach 3.5) and Purdue University (which operates at Mach 6), it would be less expensive to build a new quiet facility operating at about Mach 2 than to do the flight tests that would otherwise be required.

Plans for the Aerodynamics, Aerothermodynamics, and Plasmadynamics element of the Hypersonics Project include fundamental research on turbulence and boundary-layer physics, and one task (HYP.04.04.6) intended to demonstrate boundary flow control using a microwave plasma. However, this activity does not include improved actuators as a goal, and it is unlikely to significantly improve the aerodynamic performance of hypersonic vehicles.

A3 Novel aerodynamic configurations that enable high-performance and/or flexible multimission aircraft

SFW	SRW	Supersonics	Hypersonics	Airportal	Airspace	IVHM	IIFD	IRAC	Aging A/C

This R&T challenge has the following milestones:

- Develop a family of aircraft configurations with cruise efficiency twice as high as conventional aircraft.
- Demonstrate design approaches to develop novel configurations able to operate from small airfields.
- Validate the ability to predict the performance of novel airframe configurations using data from ground and flight tests.

The SFW Project has a substantial research effort in developing tools to predict advanced-concept airplane performance. It has defined a trade space that includes aerodynamic efficiency, noise, and emissions as key criteria to evaluate advanced configurations. Both conventional and hybrid wing families of configurations are being explored, along with powered lift and flow-control concepts to reduce minimum runway length required for takeoff and landing. In the case of the BWB high cruise efficiency configuration, these tools are being validated by flight test.

The Subsonic Rotary Wing (SRW) Project has no focused research to design or develop novel aerodynamic configurations for rotorcraft. The introduction to the project description mentions the compound slowed rotor concept, but the rest of the document does not describe any foundational or integrated research that would support this challenge by developing the concept.

The Supersonics Project is developing tools, but there is little or no effort to apply those tools to develop and predict the performance of notional aircraft configurations. Some external researchers working under NASA Research Announcements (NRAs) may be doing some analyses of their own notional aircraft configurations, but this work would be of much greater value if NASA were to define one or more notional aircraft configurations as a common reference.

A4a Aerodynamic designs and flow-control schemes to reduce aircraft and rotor noise

SFW	SRW	Supersonics	Hypersonics	Airportal	Airspace	IVHM	IIFD	IRAC	Aging A/C

This R&T challenge has the following milestones:

- Improve techniques for prediction and control of the aeroacoustics associated with high-lift devices, protuberances, and cavities for fixed-wing aircraft.

- Develop techniques for the prediction and design of quiet drag devices for fixed-wing aircraft.
- Improve understanding and modeling of unsteady fluid–structure interactions and resulting noise radiation for rotary- and fixed-wing aircraft.
- Demonstrate novel rotor system design tools that can be used to reduce rotor noise with minimum performance sacrifices for rotorcraft.

Research plans for the SFW, SRW, and Supersonics Projects support a wide array of research activities that would support the above milestones. For example, research related to noise is a large part of the Propulsion and Power Systems element and the Airframe Systems element of the SFW Project. In addition, NASA's previous participation in the Quiet Technology Demonstrator, which included partners from manufacturers and the airlines in the flight testing of advanced noise control techniques for SFW aircraft, provided a wealth of modeling, design, and validation experience. This is an important example of collaboration to facilitate technology transition to in-flight use. In addition, aeroacoustics research by the SRW Project includes foundational research to advance the understanding of sources and mechanisms of noise generation and propagation, experimental validation of predictive tools, and the stipulation of explicit metrics against which advancements in this field would be measured. The Supersonics Project is supporting variable-cycle engine research, which would help reduce noise.

A4b Accuracy of prediction of aerodynamic performance of complex 3-D configurations, including improved boundary-layer transition and turbulence models and associated design tools

SFW	SRW	Supersonics	Hypersonics	Airportal	Airspace	IVHM	IIFD	IRAC	Aging A/C

This R&T challenge has the following milestones:

- Develop improved techniques for the prediction of boundary-layer transition on 3-D configurations and validate them against ground- and flight-test data.
- Demonstrate computationally efficient techniques to couple aerodynamic and structural analysis tools.
- Develop structured techniques for predicting performance in the presence of parameter uncertainties.

The SFW Project plans to improve the ability to predict high-lift performance by developing 3-D prediction models that would be validated in wind tunnel tests with active flow-control experiments.

The SRW Project is studying structured and unstructured grids in computational fluid dynamics (CFD)-based modeling applicable to rotorcraft. The project also includes research that links these numerical models to structural and acoustic analysis capabilities. There are no plans, however, to model the effects of uncertainty in the predictive capabilities of these models or to validate the numerical models using flight tests.

The Supersonics Project is developing and validating models that address this challenge, including code development for transition and turbulence modeling of 3-D configurations. These models can and should be validated using existing data.

The Aerodynamics, Aerothermodynamics, and Plasmadynamics and Vehicle Technology Integration elements of the Hypersonics Project include many computational and experimental tasks to investigate boundary-layer transition and turbulence. The application and evaluation of the resulting models in complex 3-D configurations, however, seem to have a low priority.

A6 Aerodynamics robust to atmospheric disturbances and adverse weather conditions, including icing

SFW	SRW	Supersonics	Hypersonics	Airportal	Airspace	IVHM	IIFD	IRAC	Aging A/C

This R&T challenge has the following milestones:

- Develop and validate 3-D icing prediction tools.
- Demonstrate systems with improved spatial and temporal measurements of upstream environmental conditions.
- Develop high-bandwidth techniques to respond to and mitigate the impact of upstream environmental conditions.

This challenge is relevant to subsonic fixed-wing aircraft. However, it is not necessary for the SFW Project to address this challenge because NASA's Office of Safety is supporting worthwhile research in this area. Unfortunately, there is no substantial effort related to the specific concerns of rotorcraft. The reference document for the SRW Project describes some research to (1) develop numerical models for predicting the effects of ice accretion on lifting characteristics and (2) simultaneously develop a database of experimental results obtained in wind tunnel tests for validation purposes. However, the SRW Project is not developing models of rotorcraft behavior in wind shear of aircraft flow-field immersion, nor is it developing systems to detect and mitigate the effect of upstream environmental conditions on the safe operation of rotorcraft. Some small-scale wind tunnel tests are underway to look at ice accretion for rotorcraft, but this work is too limited to make substantial progress in addressing the above milestones.

A7a Aerodynamic configurations to leverage advantages of formation flying

SFW	SRW	Supersonics	Hypersonics	Airportal	Airspace	IVHM	IIFD	IRAC	Aging A/C

This R&T challenge has the following milestones:

- Develop improved methods to accurately predict wake vortex evolution.
- Demonstrate design tools for evaluation and optimization of multiple interacting airplanes.
- Validate models and tools for formation flying using ground and flight experiments to evaluate real atmospheric effects.

The SFW Project has no planned research to address the above milestones.

A7b Accuracy of wake vortex prediction, and vortex detection and mitigation techniques

SFW	SRW	Supersonics	Hypersonics	Airportal	Airspace	IVHM	IIFD	IRAC	Aging A/C

This R&T challenge has the following milestones:

- Develop numerical techniques to predict accurately wingtip vortex trajectory, strength, and dissipation.

- Validate numerical methods with experiments and flight testing.
- Demonstrate low-cost techniques for locating and measuring the strength of wake vortices for ground-based and aircraft-based applications.
- Integrate local weather prediction techniques into larger-scale weather models.
- Investigate aircraft designs that mitigate the strength of wake vortices.

The SFW Project is doing no work to investigate aircraft designs that mitigate the strength of wake vortices, which is the only milestone for this challenge that is related to the SFW Project.

The SRW reference document describes a focus on enhancing both structured and unstructured grid flow solvers, where one of the emphasis areas is an accurate treatment of wake vortices. Plans include code validation using experimental data. However, research plans do not include investigation of rotary-wing aircraft designs that mitigate the strength of wake vortices. This is a significant shortcoming.

A goal of the NGATS Air Traffic Management (ATM)-Airportal Project is to model and to predict wake vortex behavior to enable superdensity operations. A goal of the Coordinated Approach and Departure Operations Management element of the Airspace Project is to improve the modeling and prediction of wake vortex behavior as well as the understanding of wake vortex on airport and terminal-area capacity. The scope of this research has been reduced in recent years, in that NASA does not plan to develop new sensors, and it will rely on the National Oceanic and Atmospheric Administration and the Federal Aviation Administration (FAA) to take the lead in research related to the determination and characterization of weather and hazards such as wake vortices. However, these are not significant shortcomings in terms of the contribution of the NGATS ATM-Airportal Project to overcoming this challenge.[1]

The NGATS ATM-Airspace Project is doing no work related to in-flight applications of wake vortex research.

A9 Aerodynamic performance for vertical and short takeoff and landing (V/STOL) and extremely short takeoff and landing (ESTOL), including adequate control power

SFW	SRW	Supersonics	Hypersonics	Airportal	Airspace	IVHM	IIFD	IRAC	Aging A/C
	■								

This R&T challenge has the following milestones:

- Develop low-drag high-lift systems.
- Demonstrate systems to provide pitch trim and control power at low speeds.
- Develop new techniques for active twist control of rotors.
- Demonstrate low-cost, simple flow-control techniques for prevention of leading-edge separation from vertical and/or short takeoff and landing (V/STOL) wings.
- Improve wing design and fuselage shaping to reduce transonic cruise drag.

Plans for the SFW Project include development of technologies, such as active flow-control and morphing materials, that could enable advanced V/STOL and extremely short takeoff and landing (ESTOL) configurations, but plans do not include research to support other aspects of this challenge.

[1]Another concurrent National Research Council study is focused exclusively on NASA's wake vortex research. The results of that study were not available in time to factor them into this report.

The rotorcraft portion of this challenge is focused on active twist control of rotors to enhance aerodynamic performance at low speeds, and the SRW Project is conducting no relevant research.

A10 Techniques for reducing/mitigating sonic boom through novel aircraft shaping

SFW	SRW	Supersonics	Hypersonics	Airportal	Airspace	IVHM	IIFD	IRAC	Aging A/C

This R&T challenge has the following milestones:

- Develop guidelines for allowable exposure of the public to sonic booms.
- Develop accurate techniques for the prediction of sonic boom propagation through the atmosphere under realistic environmental conditions.
- Demonstrate novel aircraft shapes that minimize sonic boom levels.

Historically, NASA has been the leader in developing sonic boom prediction techniques. The sonic boom element of the Supersonics Project is focused on modeling sonic booms, and a more vigorous effort is needed to make substantial and timely progress in developing design evaluation tools that could be used to (1) predict the sonic boom characteristics of various aircraft shapes and (2) facilitate the design of new aircraft shapes with lower levels of sonic boom. There is no evident validation process for the computer codes being developed.

A11 Robust and efficient multidisciplinary design tools

SFW	SRW	Supersonics	Hypersonics	Airportal	Airspace	IVHM	IIFD	IRAC	Aging A/C

This R&T challenge has the following milestones:

- Develop and validate physics-based models to predict performance for novel aircraft configurations.
- Assess a family of aircraft configurations with major improvement in cruise efficiency, including a quantitative description of the benefits and risks.
- Assess novel concepts for flexible multimission aircraft, including a description of potential benefits in performance and cost.
- Conceive design approaches to develop novel V/STOL and ESTOL configurations.
- Validate design codes to predict the performance of novel airframe configurations by comparing code predictions with ground and flight tests.

The SFW Project is applying and validating tools for advanced configurations. A particular strength of this effort is the collaboration of NASA researchers with U.S. Air Force and industry researchers to validate models through flight test.

The SRW Project plans to develop a number of modeling and simulation tools, but it is not developing the integrated, multidisciplinary modeling techniques necessary to develop an MDAO capability for rotary-wing applications.

The Supersonics Project is developing appropriate tools but has not defined the configurations and tests (ground and/or flight) necessary for validation.

The Physics Based MDAO element of the Hypersonics Project includes tasks to develop multidisciplinary design capabilities in the hypersonic flow regime. However, the Hypersonics Project will not exercise these tools to investigate the performance of any specific class of aircraft configurations.

PROPULSION AND POWER

This section summarizes the committee's assessment of NASA research related to the top 10 R&T challenges involving propulsion and power (Area B) in the *Decadal Survey of Civil Aeronautics* (NRC, 2006).

B1a Quiet propulsion systems

SFW	SRW	Supersonics	Hypersonics	Airportal	Airspace	IVHM	IIFD	IRAC	Aging A/C

This R&T challenge has the following milestones:

- Develop validated physics-based models to predict engine noise and conduct trade-off studies.
- Improve understanding and prediction capabilities, and develop propulsion cycles compatible with noise and emissions reduction.
- Develop advanced low-noise fan designs, liner concepts, and active control technologies.
- Develop concepts to reduce installed noise (e.g., adaptable chevrons).
- Develop and demonstrate propulsion designs that show the feasibility of technologies to reduce noise by 10 dB (in 15 years) from Boeing 777/GE 90 levels.

The SFW Project has established the goal of reducing the noise of commercial aircraft at the three certification points (takeoff, sideline, and approach) by a total of 32 to 42 dB below current (Stage 4) standards over the next 5 to 11 years.[2] It seems unlikely that NASA will be able to achieve these ambitious goals, which are more aggressive than the goals established by the *Decadal Survey*. Nonetheless, NASA is supporting important research to reduce aircraft noise, including trade-off studies of noise versus CO_2 emissions for different engine cycles, and the assessments in this section are based on the ability of the NASA aeronautics program to address the challenges and milestones in the *Decadal Survey of Civil Aeronautics,* not NASA's own goals.

[2]For purposes of certification and regulatory compliance, the noise produced by commercial aircraft is determined by three measurements of noise on the ground, as follows:

- Takeoff noise is measured 6,500 meters from the point where the aircraft releases its brakes, as the aircraft flies overhead after takeoff.
- Sideline noise is measured at a point 450 meters to the left or the right of the runway centerline, at a point where the noise level is greatest after takeoff, as the aircraft flies past.
- Approach noise is measured at a point 2,000 meters from the threshold of the runway, as the aircraft flies overhead prior to landing.

NASA has established an "N + 1" goal of reducing noise at the above points by a cumulative total of 42 dB below the "Stage 3" standard defined by the International Civil Aviation Organization and adopted by the Federal Aviation Administration. The N + 2 goal is 52 dB below Stage 3. The current standard for type certification of new aircraft designs is Stage 4, which is 10 dB quieter than Stage 3 (cumulative, for all three certification points). Thus, the N + 1 and N + 2 goals represent a cumulative improvement of 32 to 42 dB (at all three points) relative to current limits, or about 11 to 14 dB at any one point.

This R&T challenge is targeted primarily at the prediction and mitigation of noise from jet engines. Plans for the SRW Project include research to (1) explore quiet propulsion and drive systems to increase cabin comfort and (2) improve the ability to predict noise and determine how it will propagate through he atmosphere under various conditions.

Plans for the Supersonics Project include research that effectively addresses relevant aspects of this challenge, including evaluation of empirical and statistical noise prediction tools, development of an improved statistical model for broadband shock noise, and jet noise studies.

B1b Ultraclean gas turbine combustors to reduce gaseous and particulate emissions in all flight segments

SFW	SRW	Supersonics	Hypersonics	Airportal	Airspace	IVHM	IIFD	IRAC	Aging A/C

This R&T challenge has the following milestones:

- Understand particulate matter formation mechanisms and kinetics and develop fuel additives to disrupt formation.
- Understand air toxicity measurement techniques and the impact of particulate matter on human health and welfare.
- Improve understanding and prediction capabilities and develop optimized approaches for mixing in multiphase flows.
- Develop large eddy simulations with optimized subgrid models that contain key physics needed to capture chemical reactions, mixing, and unsteady combustor phenomena.
- Develop physics-based, reduced-order combustor models, including emissions, combustion instability, blowoff, and flashback, for inclusion in intelligent engine control systems.
- Develop validated chemical mechanisms that describe fuel kinetics.
- Develop and demonstrate combustor designs that show the feasibility of technologies to reduce oxides of nitrogen (NO_x) emissions by 85 percent while also reducing particulate matter, relative to the limits set by the International Civil Aviation Organization in 1996 for future large and regional subsonic engines (with pressure ratios of 55:1 and 30:1, respectively).

The SFW and Supersonics Projects are supporting research to reduce emissions. This research encompasses all but the last of the above milestones, and NASA is providing some support, such as test facilities, to support industry-funded research. Even so, industry seems to be taking the lead in engine emissions research. Given NASA's expertise in studies of engine emissions, the state of the art in this very important area would advance more quickly if NASA were more proactive.

B3 Intelligent engines and mechanical power systems capable of self-diagnosis and reconfiguration between shop visits

SFW	SRW	Supersonics	Hypersonics	Airportal	Airspace	IVHM	IIFD	IRAC	Aging A/C
■		■							

This R&T challenge has the following milestones:

- Develop better computational simulation tools to understand operability limits.

- Develop better life prediction tools.
- Develop improved steady-state and dynamic performance checks.
- Develop improved health diagnostics systems.
- Develop new health prediction systems.
- Develop improved clearance control systems.
- Develop active compressor stall control.
- Develop active combustion control.

This R&T challenge calls for the development of computational simulation tools to understand the new generation of engines that have a high degree of self-diagnostics and regulation.

Relevant research by the SFW and Supersonics Projects is unlikely to make a significant difference to the state of the art; most of the research relevant to this challenge for these flight regimes is being funded by organizations other than NASA.

Most of the diagnostics research described in the SRW reference document pertains to rotorcraft drive systems and not the engines themselves. This research would investigate active control for increased stall margins, and it includes experimental validation of models.

The Integrated Vehicle Health Management (IVHM) Project is addressing some aspects of this challenge, such as the issue of monitoring propulsion systems with smaller unique high-temperature sensors in the gas path and the development of diagnostic systems to predict engine health.

B4 Improved propulsion system fuel economy

SFW	SRW	Supersonics	Hypersonics	Airportal	Airspace	IVHM	IIFD	IRAC	Aging A/C

This R&T challenge has the following milestones:

- Demonstrate laboratory-scale materials for 1500°F compressor disks.
- Demonstrate materials for full-scale, 1500°F compressor disks.
- Perform 1,000-hour test of a 50-horsepower per pound speed reduction gearbox.
- Test a reduced-weight, high-bypass-ratio engine and nacelle-to-wing configuration in a wind tunnel.
- Demonstrate an acceptably low-cost, advanced high-pressure turbine cooling system.

The Durability of Engine Superalloy Disks element of the Aging Aircraft and Durability Project is working to improve the durability of disks at operating temperatures up to 1300°F, but this challenge calls for research to enable operating temperatures up to 1500°F (for the purpose of improving fuel economy), and research to develop new materials able to function at 1300°F is likely to focus on classes of materials that will be unsuitable for operations at 1500°F. As a result, this research is not relevant to this challenge.

Neither the SFW Project nor the Supersonics Project is working on developing compressor disks with an operating temperature of 1500°F, higher-pressure-ratio engines, or the thermal problems associated with those engines. However, the SFW Project is working to improve the engine component efficiency and aircraft aerodynamics, which should improve specific fuel consumption.

The SRW Project is addressing this challenge through fundamental research in oil-free engines for rotary-wing flight and the integration of such engines with an optimized drive system gearbox. However, no significant full-scale demonstrations are planned. Also, the *Decadal Survey* predicts that, over the long term, research related to this challenge should be able to improve fuel economy by 30 percent relative to the GE 90 (for large commercial engines) and by 30 percent relative to T700/CT7 (for small engines). However, the SRW Project has not set specific targets for improving fuel efficiency.

B5 Propulsion systems for short takeoff and vertical lift

SFW	SRW	Supersonics	Hypersonics	Airportal	Airspace	IVHM	IIFD	IRAC	Aging A/C

This R&T challenge has the following milestones:

- Demonstrate pressure ratios between 25:1 and 30:1 and turbine inlet temperatures of 2800°F for 3,000-shaft-horsepower-class engine components.
- Develop and validate the design tools required for candidate gearboxes and clutch systems.
- Demonstrate highly reliable gearboxes, which have transfer efficiencies of about 99.8 percent and power-to-weight ratios of about 50 horsepower per pound.
- Demonstrate clutch system technologies with 10,000-cycle life and a probability of failure of 1×10^{-6} over the life of the system.

The SFW Project has no planned research to address the above milestones.

The SRW Project includes research on optimization of gearbox-engine systems, wide-operability engines, and efficient, high-power-density engine technologies. However, these tasks only use metrics related to the accuracy of tools, and they do not target specific levels of propulsive efficiency. Neither the SFW Project nor the SRW Project includes thrust vectoring or engine-assisted lift research.

B6a Variable-cycle engines to expand the operating envelope

SFW	SRW	Supersonics	Hypersonics	Airportal	Airspace	IVHM	IIFD	IRAC	Aging A/C

This R&T challenge has the following milestones:

- Develop variable exhaust nozzle technology to optimize fuel burn.
- Develop improved thermal management systems.
- Develop ceramic matrix composite technologies for hot section components.
- Develop highly loaded, high-speed bearings.
- Develop probabilistic analysis for more accurate designs and life prediction.
- Develop improved turbine cooling technology.
- Develop high-temperature combustors to accommodate increased operating pressure ratios.
- Develop improved aircraft-engine integration tools.

The SFW Project is supporting some component research that addresses most of the above milestones, but the U.S. Air Force is leading research in this area.

Plans for the Supersonics Project include research that would likely make substantial progress in the near term in meeting the intent of this R&T challenge, although the usefulness of this work will be limited somewhat by the lack of any notional vehicle to guide the research.

B6b Integrated power and thermal management systems

SFW	SRW	Supersonics	Hypersonics	Airportal	Airspace	IVHM	IIFD	IRAC	Aging A/C
■	■	■	■						

This R&T challenge has the following milestones:

- Identify and mature new business models for the design, development, validation, and support of hardware and software components of integrated systems.
- Develop an object-oriented modeling infrastructure that allows networking resources to operate across different hardware platforms and geographic sites.
- Develop new engine-airframe systems integration architectures for both subsonic and higher-speed flight.
- Develop physics-based subsystem component models that can analyze transient operations.
- Develop and mature concepts for the integration of fuel cell technology as secondary power sources.
- Develop advanced electric or electromechanical actuators that have rapid response, high power-to-weight, and low heat rejection.
- Develop subsystem components that can survive in more stressful thermal environments, require less cooling, and reject less waste heat, including thermally efficient fuel pumps and high-temperature electronics for power management and distribution systems.
- Develop lightweight, high-energy-density batteries.
- Develop advanced heat exchanger technologies.

The high velocity of hypersonic vehicles make it particularly important to manage the power and thermal balance. However, thermal management is important for aircraft in all flight regimes. Even so, neither the Hypersonics, Supersonics, SFW, nor SRW Project has planned research to address the above milestones.

B8 Propulsion systems for supersonic flight

SFW	SRW	Supersonics	Hypersonics	Airportal	Airspace	IVHM	IIFD	IRAC	Aging A/C
		■							

This R&T challenge has the following milestones:

- Establish needed boundary conditions, initial conditions, and other inputs and outputs for each module of multidisciplinary, system-level design tools.
- Develop technology that will enable supersonic aircraft to meet Stage 4 noise standards.
- Validate boundary-layer control techniques for inlet performance and drag reduction.
- Demonstrate a supersonic variable-cycle engine with specific fuel consumption of 1.1 or lower and a thrust-to-weight ratio of at least 6.
- Demonstrate high-performance, low-drag, noncircular inlet designs.
- Obtain flight-test data on noise, emissions, human annoyance caused by sonic boom, and system interactions across the flight regime.

The Supersonics Project is not working to achieve a specific fuel consumption of 1.1 or a thrust-to-weight ratio of at least 6. Little or no work is being done to achieve the higher pressure ratios, higher operating temperatures, extended flight times, and long times between major maintenance that the propulsion systems for a commercially viable supersonic transport will need to demonstrate.

B9 High-reliability, high-performance, and high-power-density aircraft electric power systems

SFW	SRW	Supersonics	Hypersonics	Airportal	Airspace	IVHM	IIFD	IRAC	Aging A/C

This R&T challenge has the following milestones:

- Demonstrate tenfold increase in power density for suitable electric generators and motors.
- Demonstrate fivefold increase in energy and power density of suitable batteries and hybrid storage systems (e.g., the battery–ultracapacitor).
- Demonstrate an order-of-magnitude lighter optimized power system architecture (including, for example, a direct-current power bus, remotely controlled loads, and a wireless system control).
- Demonstrate intelligent power management and distribution using advanced system models and wireless sensors or sensorless control technologies for graceful degradation and failsafe operation.
- Demonstrate advanced analysis and simulation tools for multiconverter power systems, which can predict new modes of system dynamics and instability.

Neither the Supersonics, SFW, nor SRW Project has planned research to address the above milestones.

The IVHM Project is developing the Advanced Diagnostics and Prognostics Testbed facility at NASA Ames Research Center. This system will investigate intelligent power management systems for graceful and failsafe degradation and develop modeling and simulation tools that can assess power systems dynamics. However, the IVHM Project is doing no work to increase the energy and/or power density of electric generators, motors, batteries, hybrid power storage systems, or power system architectures.

B10 Combined-cycle hypersonic propulsion systems with mode transition

SFW	SRW	Supersonics	Hypersonics	Airportal	Airspace	IVHM	IIFD	IRAC	Aging A/C

This R&T challenge has the following milestones:

- Develop advanced diagnostics capable of measuring time-averaged and time-resolved flow parameters and their correlations.
- Demonstrate ramjet-scramjet (dual mode) transition and isolator performance for a simplified geometry with alternately clean and vitiated air.
- Conduct transient experiments to simulate cowl door movement for turbine-ramjet mode transition and cowl lip movement to control inlet contraction.
- Demonstrate injection, mixing, and combustion using simple fuel injectors and with alternately clean and vitiated air.
- Conduct inlet studies with variable angles of attack and sideslip angles.

- Investigate new engine configurations using inward-turning inlets, elliptical cross sections, and so on.

The Propulsion and Propulsion Technology Integration elements of the Hypersonics Project are conducting a significant amount of experimental and computational work related to the development of highly reusable and reliable launch systems. These systems would use hypersonic vehicles powered by air-breathing, combined-cycle, propulsion engines, and a substantial amount of related work supported by the Hypersonics Project would meet the intent of this R&T challenge.

MATERIALS AND STRUCTURES

This section summarizes the committee's assessment of NASA research related to the top 10 R&T challenges involving structures and materials (Area C) in the *Decadal Survey of Civil Aeronautics* (NRC, 2006).

C1 Integrated vehicle health management

SFW	SRW	Supersonics	Hypersonics	Airportal	Airspace	IVHM	IIFD	IRAC	Aging A/C

This R&T challenge has the following milestones:

- Develop lightweight sensor networks that characterize the state of materials and structures over large areas.
- Develop very-low-power or self-powered wireless sensors capable of operation in harsh environments.
- Develop artificial intelligence to automatically assess structural integrity from sensor responses and implement damage mitigation protocols.
- Develop components and sensors that are cost-competitive and available from multiple vendors.
- Flight test full-scale IVHM systems to detect multisite damage.

This challenge is unique in that its scope encompasses almost all the work being conducted by a single project: the IVHM Project. The discussion of this R&T challenge in the *Decadal Survey of Civil Aeronautics* notes that "with a national fleet of aging aircraft and infrastructure in an industry with low profit margins, IVHM is increasingly important. . . . Early detection of impending failures in aircraft materials, structures, and wiring is critical for avoiding fatalities as a part of the aging aircraft program. IVHM also reduces time lost to scheduled maintenance and reduces the likelihood of unscheduled downtime" (NRC, 2006, p. 113). In addition, the reference document for the Aging Aircraft and Durability Project states that that "aircraft aging is a significant national issue" (NASA, 2006, p. 1).

The average age of U.S. commercial aircraft dropped in the aftermath of the attacks on September 11, 2001, as many older aircraft were retired. On a global basis, the average age of regional jets operated by scheduled airlines is now less than 7 years. By contrast, the average age of narrowbody jets operated by scheduled airlines, nonscheduled airlines, and air cargo carriers is 13.5 years, 21.5 years, and 28.2 years, respectively (OAG, 2007). In addition, the average age of aircraft operated by the Department of Defense (DoD) is also increasing, sometimes to the point of exceeding the original design service life of the vehicles.

Even though aging is an issue with existing commercial and military aircraft, the project description for the Aging Aircraft and Durability Project says that its focus "is on aging and damage processes in 'young' aircraft, rather than life extension of legacy vehicles" (NASA, 2006, p. 1). NASA's research plans in this area emphasize new and emerging material systems and fabrication techniques and the potential hazards associated with aging-related degradation. Such a focus would greatly limit the ability of the project to have any impact on the air transportation system in the near or mid term.

Advanced IVHM can expedite the introduction of innovative material systems and structural concepts in rotorcraft. However, the reference document for the SRW Project does not address research related to the design or development of rotorcraft IVHM technologies or systems. In particular, the SRW Project does not focus on the use of sensor networks to gage the real-time state of rotorcraft systems, nor does it support research on the use of data to mitigate the influence of structural or material degradation.

The IVHM Project is developing health management sensors and sensor networks applicable to airframes, aircraft systems, propulsion systems, and environmental hazards (such as ice, radiation, and lightning strikes). This work includes fiber-optic sensors for determining the distribution of strain and temperature in an airframe system, high-temperature silicon-carbide sensors, extensive studies of the impact of ionization and radiation, data mining research. NASA's IVHM research on ice as a hazard does not clearly add value to icing research done elsewhere. However, IVHM research on data mining is novel and has the possibility of creating new paradigms for aircraft maintenance over the life of the vehicle. NASA may be able to create universal formats, protocols, and databases for representing data and informatics about any air vehicle. NASA is also developing a testbed at Ames Research Center that simulates vehicle power systems and allows IVHM principles to be applied to power network controllers and systems. This testbed seems relevant to NASA's future spacecraft missions, however, and it is uncertain how it can be used for direct support of aircraft.

The Aging Aircraft and Durability Project has three research areas at the multidisciplinary capabilities level (i.e., Level 3, as defined by ARMD). These research areas are Detect, Predict, and Mitigate. Some of the research related to the Detect area supports this challenge and some of the above milestones, although in-flight monitoring is not identified as a goal. Reduction of maintenance downtime is one goal of this challenge, but NASA staff working on the Aging Aircraft and Durability Project lack firsthand experience with commercial transport maintenance operations. The Aging Aircraft and Durability Project's low level of effort and broad scope bring into question its ability to make significant progress.

C2 Adaptive materials and morphing structures

SFW	SRW	Supersonics	Hypersonics	Airportal	Airspace	IVHM	IIFD	IRAC	Aging A/C

This R&T challenge has the following milestones:

- Identify new morphing missions and designs for reconfigurable civil aircraft, including supersonic aircraft with low sonic boom.
- Develop the next generation of high-strain, adaptive materials or devices that can be activated and deactivated for repositioning, with actuation deformation up to 100 percent.
- Develop novel integrated adaptive materials that allow wing surfaces and fuselages (including inlets) to rapidly change shape or alter load paths.
- Conduct scaled wind tunnel and flight tests on active, morphing aircraft to enable innovative, lightweight designs.

- Develop new, structurally integrated adaptive devices for flow control on a commercial aircraft to, for example, reduce drag and improve performance in off-design conditions.
- Develop analysis and design tools that account for and accurately predict nonlinear behaviors of adaptive materials and morphing structures.

The SFW Project is doing good work in basic material composition and adaptive structures and materials, although a greater effort to apply this research to structural configurations would be necessary to fully meet the intent of this R&T challenge.

The SRW Project includes little or no research on the use of adaptive materials for rotorcraft structural applications.

The Supersonics Project is concentrating on aeroelastic issues, but not on adaptive materials or morphing.

C3 Multidisciplinary analysis, design, and optimization

SFW	SRW	Supersonics	Hypersonics	Airportal	Airspace	IVHM	IIFD	IRAC	Aging A/C

This R&T challenge has the following milestones:

- Develop multidisciplinary analysis tools that incorporate aerodynamics, structural dynamics, vibration, thermal response, acoustic response with structural response to mechanical loads.
- Extend multidisciplinary tools to incorporate explicit mathematical modeling of design issues such as manufacturing processes, life-cycle cost, and repairability.
- Develop efficient approaches for multivariable optimization.
- Develop efficient and effective search processes for the analysis of large, complex systems.
- Develop approaches for modeling uncertainty in data-lean environments.
- Develop computationally efficient methods for reliability assessment.
- Develop a systematic approach for modeling risk and uncertainty in complex coupled systems.

The SFW and Supersonics Projects are demonstrating the implementation of relevant tools in an integrated program that addresses the above milestones. For example, the Computationally Guided Multifunctional Materials Development element of the SFW Project includes development of physics-based models to increase the fundamental understanding of materials science, establish the interconnectivity of the multiple disciplines required to provide parametric analysis tools, and develop examples of key structure-property relationships for novel materials. Validation would be achieved by experiment. Ultimately, this approach to materials design would enable the tailoring of key properties such as strength, durability, acoustic damping, and conductivity to address multifunctional applications.

The SRW Project is developing a suite of analysis tools, but it is not supporting substantial work in assembling the tools into an MDAO capability for rotary-wing applications. The SRW Project is not explicitly examining the computational fidelity of these tools for different stages of the MDAO process, nor is it making a significant effort in integrating the various analysis tools and linking them to a design optimization capability. Neither is the SRW Project quantifying or representing the influences on uncertainty in the analysis and design tools.

Several elements of the Hypersonics Project address this R&T challenge. The Materials and Structures element includes milestones related to multidisciplinary thermal and structural analysis methods for the design and evaluation of hypersonic airframe and propulsion structures that incorporate life prediction and reliability models. The Vehicle Technology Integration element includes milestones related to the development of tools and methodologies to design and analyze the highly coupled multidisciplinary systems problems presented at the vehicle system level in support of hypersonic vehicles. The Physics Based MDAO element includes milestones related to the development of advanced physics-based multidisciplinary predictive design, analysis, and optimization tools focused on supporting Highly Reliable Reusable Launch Systems (HRRLS) and High Mass Mars Entry Systems (HMMES) mission vehicles.

The Aging Aircraft and Durability Project is organized around eight challenge problems, each of which is more or less a stand-alone research effort. They may touch on some of the above milestones, but the project details are not integrated to produce multidisciplinary design, analysis, or optimization. A systematic approach for modeling risk and uncertainty in complex coupled systems (the last milestone) would be very helpful in identifying and prioritizing aging aircraft and durability issues.

C4 Next-generation polymers and composites

SFW	SRW	Supersonics	Hypersonics	Airportal	Airspace	IVHM	IIFD	IRAC	Aging A/C

This R&T challenge has the following milestones:

- Demonstrate fabrication of composites with many different types of reinforcement fibers.
- Integrate adaptive materials to increase functionality.
- Develop techniques for manufacturing, processing, and dispersion of nanoscale reinforcements.
- Develop a fundamental understanding of how different kinds of reinforcements (e.g., nano, functional, or hybrid additives) affect the performance of polymers and composites.
- Improve damage tolerance for high-temperature polymers.
- Develop effective life prediction models for polymers and composites.
- Investigate environment-friendly end-of-life reuse or disposal strategies.

The Materials, Structures, and Mechanical Components element of the SFW Project is addressing a number of composites issues, but a more active program is needed in this very important area.

The SRW Project has no planned research to address milestones relevant to rotorcraft.

Plans for the Supersonics Project include research that would make substantial progress in the near term in meeting the intent of this R&T challenge.

Three of the eight challenge problems (CPs) being addressed by the Aging Aircraft and Durability Project address polymers and composites, as follows:

- CP-03: Durability and Structural Integrity of Composite Skin-Stringer Fuselage Structure
- CP-04: Durable Bonded Joints
- CP-05: Durability of Engine Fan Containment Structure

The challenge problem on Durability and Structural Integrity of Composite Skin-Stringer Fuselage Structure (CP-03) directly addresses this challenge, but given the large level of effort in industry and extensive role that composites are already playing in commercial products, the timing of this effort and the low level of effort will limit the value of this research. The value of CP-04 and CP-05 is uncertain given the low levels of effort in both areas. In addition, for CP-05, blade containment systems using current technology already demonstrate a high level of protection, especially for composite blades.

C5 Noise prediction and suppression

SFW	SRW	Supersonics	Hypersonics	Airportal	Airspace	IVHM	IIFD	IRAC	Aging A/C

This R&T challenge has the following milestones:

- Measure noise signatures in controlled environments such as anechoic wind tunnels, for a range of flight conditions.
- Predict noise signatures using advanced multidisciplinary methodologies, validating against test data for level and maneuvering flight modes.
- Develop efficient structural solutions for interior noise control, that is, structural optimization.
- Design non-load-bearing passive noise control.
- Design active controls for interior and exterior noise through smart structures technology.
- Develop low-noise rotors.
- Selectively flight test full-scale systems with noise signature measurement.

The SFW Project is supporting research on fundamental noise control to alleviate structural noise. The Systems Analysis, Design, and Optimization element recognizes noise as an important aspect of MDAO modeling, although the principal investigator (PI) reports that there is a shortage of noise analysts available to support this work.

The Supersonics Project has no planned research to address the above milestones.

The *Decadal Survey of Civil Aeronautics* notes that noise prediction and suppression are very important for rotorcraft. The reference document for the SRW Project emphasizes the need for research to reduce both interior and exterior noise, and it characterizes noise generation as a problem involving coupled structures, fluid dynamics, and acoustics. The use of smart materials to reduce interior noise is one task (among many) in the aeroacoustics area of the SRW Project. NASA is a key player in DARPA's Helicopter Quieting Program,[3] although that is not mentioned in the SRW Project's reference document.

C6a Innovative high-temperature metals and environmental coatings

SFW	SRW	Supersonics	Hypersonics	Airportal	Airspace	IVHM	IIFD	IRAC	Aging A/C

This R&T challenge has the following milestones:

- Define required models and a model integration strategy to provide necessary functionality for simulations.
- Select models for further development, based in part on how well they are aligned with materials systems that provide the greatest benefit for propulsion systems.
- Develop models for selected substrates and associated environmental coatings; determine all the physical parameters required by the models.

[3]See <www.darpa.mil/ucar/text/programs/hq.htm>.

- Validate the models by applying them to the development of new materials that are selected in concert with industry.

The SFW Project has no planned research to address the above milestones.

The reference document for the SRW Project affirms the importance of advanced material concepts for improved efficiency of small rotorcraft engines. Plans for the SRW Project include research into monolithic silicon nitride for hot sections and silicon carbide coatings. As suggested in the *Decadal Survey,* validation of modeling through experimental tests in the high-temperature engine test facilities is also planned.

The Supersonics Project is supporting some research on modeling and development of high-temperature materials, coatings, and processes, but it is not supporting research related to other aspects of this challenge.

The Materials and Structures element of the Hypersonics Project includes experimental tests of metallic materials and structures, but relevant modeling efforts do not seem to be part of this project.

In the Aging Aircraft and Durability Project, the first challenge problem (CP-01: Damage Methodology for Metallic Airframe Structures) and the second challenge problem (CP-02: Structural Integrity of Integral Metallic Structure) do not specifically address metal high temperatures or coatings, but the results of this research could contribute to overcoming this challenge, particularly with regard to a better understanding of aluminum structures in elevated temperature environments. Plans for CP-06 (Durability of Engine Superalloy Disks) and CP-07 (Durability of Engine Hot Section) address this challenge, and research in this area clearly could have significant value, but the low planned levels of effort (2.9 labor-years for CP-06 and 1.5 labor years for CP-07) draw into question the likelihood that planned work will substantially advance the state of the art.

C6b Innovative load suppression, and vibration and aeromechanical stability control

SFW	SRW	Supersonics	Hypersonics	Airportal	Airspace	IVHM	IIFD	IRAC	Aging A/C

This R&T challenge has the following milestones:

- Predict vibration using advanced CFD methodologies and validate experimentally.
- Predict aeromechanical stability for advanced configurations and expanded flight envelope (including hypersonic flight) and validate experimentally.
- Measure vibratory loads and vibration signatures under controlled, wind tunnel environments for a range of flight conditions.
- Develop novel techniques for control-oriented modeling.
- Selectively flight test full-scale systems, measuring vibration signatures and damping levels at level and maneuvering flight conditions.
- Innovate and employ active or passive techniques to minimize vibration and increase stability margin.
- Develop MDAO techniques to develop low-vibration, stable systems.

The SFW Project is doing substantial work on aspects of this challenge related to the engine and airframe.

The *Decadal Survey of Civil Aeronautics* emphasizes the importance of load suppression, vibration reduction, and aeromechanical stability control to rotorcraft. The reference document for the SRW

Project describes specific research tasks that would address these areas by developing and validating predictive tools, using experimental test facilities as needed. There is a strong focus on coupling CFD codes and computational structural dynamics codes to develop predictive capabilities and to validate these predictions through wind tunnel tests. The SRW Project does not plan to conduct the full-scale tests included in the milestones for this challenge.

The Supersonics Project initially emphasizes aeroelastic approaches and controls for reducing gust loads and aeroelastic amplitudes as well as control strategies to eliminate low-frequency structural vibrations. The reduction of flutter and dynamic stresses is proposed for fiscal years (FY) 2010-2011. All of this is through the use of computer modeling. It is not clear how these techniques will be validated.

The Hypersonics Project has no planned research to address the above milestones.

C8 Structural innovations for high-speed rotorcraft

SFW	SRW	Supersonics	Hypersonics	Airportal	Airspace	IVHM	IIFD	IRAC	Aging A/C

This R&T challenge has the following milestones:

- Develop comprehensive aeromechanic analyses for high-speed rotorcraft that include tilt-rotor, tandem-rotor compound, and compound coaxial rotors for level and maneuvering flight conditions.
- Develop aeromechanics and technology tools for the drive-train system and other key components necessary for variable-speed rotors.
- Design and develop lightweight, crash-absorbing composite airframes.
- Develop technology for all-weather rotorcraft operation.
- Develop advanced composites with high damage tolerance for use in large dynamic structural components.
- Reduce required shaft power by 15 percent from current levels using elastically tailored composite blades, active structural and flow control, and advanced airfoils.
- Reduce life-cycle cost using health and usage management systems, low-cost tailored airframes, and lightweight low-vibration rotors.

The SRW Project includes specific tasks to improve the integration and design of drive-train systems, with an emphasis on variable and multispeed drive system technologies. It also includes ongoing research related to structural design for enhanced crashworthiness in rotorcraft structures. However, the SRW Project's reference document does not address (1) development of tools for innovative new rotorcraft configurations (tilt-rotors, tandem compound rotors, coaxial compound rotors) or (2) reduction of life-cycle costs using health and usage monitoring of rotorcraft structures.

C9 High-temperature ceramics and coatings

SFW	SRW	Supersonics	Hypersonics	Airportal	Airspace	IVHM	IIFD	IRAC	Aging A/C

This R&T challenge has the following near-term milestones:

- Generate material property databases appropriate for design of a high-temperature ceramic component.

- Complete full-scale testing of at least one ceramic composite component with improved performance for subsonic aircraft applications (e.g., fairing heat shields, combustor liner, or turbine airfoil).
- Develop models to optimize a structure for a new, rather than an existing, platform.
- Model crack growth under actual operating conditions.
- Develop advanced ceramic composites for large surfaces and leading-edge components for supersonic and hypersonic vehicles and complete relevant environmental testing of subcomponents.

This R&T challenge also has the following far-term milestones:

- Flight test at least one ceramic composite component for improved subsonic flight vehicles and transfer the technology to industry.
- Verify model predictions of performance using flight-test data.
- Extend model predictions to new flight speed regimes to optimize supersonic and hypersonic vehicle designs for hot structures and engine components.
- Demonstrate, through full-scale testing, at least one ceramic composite component for a supersonic or hypersonic platform.

The ceramics research by the SFW Project is inconsistent with the above milestones and will not provide timely results, particularly with respect to high-temperature requirements. NASA should consider teaming with the Air Force in this arena.

The SRW Project plans to support some research in ceramics (in the form of ceramic matrix composites). However, the SRW Project will not address the need for (1) more fabrication and testing experience with these materials and (2) a better understanding of their long-term behavior.

The Supersonics Project includes development of metal/polymer composites in the near term. The development of ceramic matrix composites and ceramic barrier coatings is delayed until FY 2011.

The Hypersonics Project in its Materials and Structures element contains many tasks for evaluation and modeling of materials and structures constructed from ceramic matrix composites for propulsion and airframe applications, but the Hypersonics Project does not include flight testing to validate these structures and materials.

In the Aging Aircraft and Durability Project, CP-06 (Durability of Engine Superalloy Disks) and CP-07 (Durability of Engine Hot Section) address coatings relevant to this challenge, but at a low level of effort (2.9 labor-years for CP-06 and 1.5 labor years for CP-07) that limits the likelihood of producing significant and timely results. Furthermore, plans for the Aging Aircraft and Durability Project do not address the need for (1) more fabrication and testing experience with these materials and (2) a better understanding of their long-term behavior.

C10 Multifunctional materials

SFW	SRW	Supersonics	Hypersonics	Airportal	Airspace	IVHM	IIFD	IRAC	Aging A/C

This R&T challenge has the following milestones:

- Develop a comprehensive analysis to predict the performance of selected monolithic and composite multifunctional materials.

- Use this analysis to guide parametric studies to explore and optimize material response with the goal of understanding the combined response of the multifunctional material.
- Fabricate materials according to model predictions.
- Evaluate material performance, both coupled and structural, and compare with analytical predictions.
- Integrate multifunctional materials into a structural component for benchtop verification.
- Conduct flight tests on a structural component.

The SFW Project has a solid plan that addresses the above milestones. Progress reported to date, however, is inconsistent with the plan.

Neither the SRW Project nor the Supersonics Project has planned research to address the above milestones.

The Materials and Structures element of the Hypersonics Project contains tasks for design and evaluation of multifunctional ablator materials. The task appears limited in scope to one type of composite material, and the project does not include any flight testing of these structures and materials for validation as specified in the R&T challenge.

The IVHM Project is supporting research to develop self-healing materials. The research supports some aspects of the R&T challenge. The integration of sensors into multifunctional materials would help develop structures capable of prognostics. The development of energy harvesting systems to collect thermal energy from the propulsion system is a good application of thermoelectric materials.

DYNAMICS, NAVIGATION, AND CONTROL, AND AVIONICS

This section summarizes the committee's assessment of NASA research related to the top 10 R&T challenges involving dynamics, navigation, control, and avionics (Area D) in the *Decadal Survey of Civil Aeronautics* (NRC, 2006).

D1 Advanced guidance systems

SFW	SRW	Supersonics	Hypersonics	Airportal	Airspace	IVHM	IIFD	IRAC	Aging A/C
				■	■				

This R&T challenge has the following milestones:

- Develop advanced algorithms and avionics for collision, terrain, and wake vortex avoidance; formation flight and cooperative and multiaircraft guidance; and ground operations guidance (taxi, takeoff, rollout, and turnoff).
- Expand facilities and programs capable of maturing the above technologies to flight-ready systems.
- Develop and adopt regulations for the certification and operation of autonomous unmanned air vehicles (UAVs) in civil airspace.

The reference document for the SRW Project includes goals related to the development of algorithms to avoid flight into terrain, particularly in all-weather operations. Plans for the SRW Project include research on precision guidance, navigation, and control capabilities for rotorcraft, although the reference document states that limited resources will constrain related research to aspects of the program that provide an integrated solution to handling qualities and dynamics problems.

The Safe and Efficient Surface Operations element of the NGATS ATM-Airportal Project explores optimization of surface routing and the roles of pilots, controllers, and systems in surface operations. None of the other research planned by either the NGATS ATM-Airportal Project or the NGATS ATM-Airspace Project addresses the above milestones (or similar milestones for challenge D10 on the safe operation of UAVs in the national airspace).

The Adaptive Intelligent Information Management task of the Integrated Intelligent Flight Deck (IIFD) Project is addressing advanced onboard guidance. However, most of the onboard guidance research supported by ARMD is conducted by the NGATS ATM-Airspace Project. For example, this research includes the generation of aircraft trajectories using algorithmic methods. The IIFD Project is conducting research to improve pilots' situational awareness by providing them with information about the trajectories of aircraft in their aircraft's critical volume, so they can make appropriate spacing decisions.

The IIFD Project anticipates providing its results to various government organizations, including the DoD. However, the IIFD Project has not considered the advances that DoD has made in integrated display systems in its most recent fighter aircraft, the F-22 and F-35. These systems integrate information from multitudes of sensors on the aircraft, on the ground, and on other aircraft. This directly relates precisely to issues that the IIFD Project is addressing, such as information quality and uncertainty, decluttering, and general spatial integration of information.

D2 Distributed decision making, decision making under uncertainty, and flight-path planning and prediction

SFW	SRW	Supersonics	Hypersonics	Airportal	Airspace	IVHM	IIFD	IRAC	Aging A/C

This R&T challenge has the following milestones:

- Develop simulation capabilities for evaluation and demonstration of certain high-performing strategies in the execution of realistic system architectures and applications.
- Develop a requirements flowdown to all affected aircraft systems, such as advanced communications, navigation, and surveillance systems.
- Develop improved, automated logic and processes for contingency management.
- Develop a methodology to support verification and validation of future systems technologies developed by this challenge.

Although the SRW Project includes a system-level milestone on Control Theory, Intelligent/Autonomous Systems, Information Processing, and Modeling, related research does not relate to the fundamental thrust of this R&T challenge as it relates to rotorcraft, which is to add autonomy and automatic controls to improve integration of rotorcraft in the air traffic management system.

The Safe and Efficient Surface Operations element of the NGATS ATM-Airportal Project explores optimization of surface routing and the roles of pilots, controllers, and systems in surface operations. The role of people in decision making is addressed in the Airportal Transition and Integration Management element, which is developing a human performance model for the roles of pilots, controllers, airport ground personnel, and others. This effort is examining the allocation of roles between humans and automation. In addition, the Coordinated Approach/Departure Operations Management element of the NGATS ATM-Airportal Project is examining Required Navigation Performance and four dimensional (4-D) trajectories for improved runway throughput.

The Traffic Flow Management, Super Density Operation, and Trajectory Prediction, Synthesis, and Uncertainty elements of the NGATS ATM-Airspace Project address the first and third milestones for this R&T challenge (see above). This research is also well aligned to the stated needs of the NextGen System (as of September 2007). The planned research should advance the state of the art, but timely undertaking of substantive research by the Super Density Operations and Trajectory Prediction elements is compromised by a shortage of qualified and experienced staff.

The IIFD Project is addressing important elements of this R&T challenge, but the research would be more effective if it did a better job of taking into account work that DoD has performed in this area. (See discussion under D1, above.)

D3 Aerodynamics and vehicle dynamics via closed-loop flow control

SFW	SRW	Supersonics	Hypersonics	Airportal	Airspace	IVHM	IIFD	IRAC	Aging A/C

This R&T challenge has the following milestones:

- Develop simpler representations of the aircraft system dynamics for control design.
- Develop distributed control algorithms and architectures.
- Demonstrate the ability to numerically solve distributed control algorithms at the Reynolds numbers associated with manned aircraft flight to demonstrate control performance.
- Implement integrated, distributed closed-loop flow-control systems.
- Design and develop lightweight, mechanized, shape-changing structures.
- Experimentally verify the performance of shape-changing aerodynamic structures before flight testing.

The SFW Project includes research and flight testing of shape-changing structures, but it does not include substantial work on the other elements of this challenge.

The focus of this challenge is on using distributed sensors and actuators to provide flow control over a wide range of flight operations. The SRW Project does not effectively address this goal.

The Supersonics Project includes research related to this challenge, but it is focused mostly on developing and testing computational codes, and it is not applying those tools to develop and predict the performance of notional aircraft configurations.

D4 Intelligent and adaptive flight control techniques

SFW	SRW	Supersonics	Hypersonics	Airportal	Airspace	IVHM	IIFD	IRAC	Aging A/C

This R&T challenge has the following milestones:

- Develop an adaptive, intelligent, fully integrated vehicle management system that can operate safely without reliable sensor information.
- Demonstrate a mature methodology for designing and analyzing flight control laws for aircraft with large numbers of highly distributed control actuators and sensors—for example, shape memory alloys and piezoelectrics.
- Demonstrate a mature methodology for using information of different degrees of reliability

without compromising flight safety (e.g., using data from what would traditionally be considered non-flight-critical systems within an inner control loop).
- Demonstrate long-term learning so that adaptation would only need to be used in novel situations. For example, following damage, the system adapts the first time it enters a particular part of the flight envelope but does not need to readapt if it leaves that part of the envelope and returns.
- Validate complex nonlinear systems to seek out worst-case scenarios that may not be identified with exhaustive testing.

Intelligent and adaptive flight control techniques require that vehicle systems be highly integrated. The reference document for the SRW Project describes research related to this challenge, but it does not directly address the above milestones. The most important rotorcraft-unique aspect of this challenge is the first milestone, and the plans for the SRW Project includes research that supports that milestone.

The Integrated Dynamics and Flight Control, Integrated Propulsion Control and Dynamics, and Airframe and Structural Dynamics elements of the Integrated Resilient Aircraft Control (IRAC) Project address key aspects of this challenge for fixed-wing applications. The research in question will investigate the integration of (1) engine and actuator health data and (2) structural degradation and damage. This research would also integrate propulsive and aerodynamic controls to optimize flight performance in damaged and degraded conditions. NASA assets such as simulation facilities, the scale-model transport testbed, and aircraft at Dryden Flight Research Center would be used to support full-scale validation of the concepts produced by this research.

D5 Fault-tolerant and integrated vehicle health management systems

SFW	SRW	Supersonics	Hypersonics	Airportal	Airspace	IVHM	IIFD	IRAC	Aging A/C

This R&T challenge has the following milestones:

- Specify nominal models and model behavior, interface, and test requirements for component and integrated system capability affected by degraded or failed operation of a representative subset of avionics and flight system components. Define suitable thresholds for levels of degraded and failed operation for component-level and system-level operations.
- Work with aircraft subsystem and flight system vendors to specify parameters that are candidates for maintenance logging. Develop models and compact representations that can incorporate measurements of these parameters in near real time and develop thresholds that can be used for on-demand maintenance activities.
- Evaluate component capability in a simulated environment (ground test and hardware in the loop). That is, take a particular subsystem, such as a real landing gear system that has been represented by an appropriate behavioral model as specified, and insert simulated faults to test for proper operation of the health monitoring system. Perform these tests for all representative subsystems that were specified above.
- Evaluate integrated system capability in a simulated environment. Take the subsystem health models previously specified and insert faults, preferably ones that were not detected as quickly as necessary by the individual component models that were evaluated in the above set of tests, and use the system models in order to evaluate the efficacy of their integrated operation.
- Test component and integrated system capability in a flight environment.

The IVHM, IIFD, and IRAC Projects are generally addressing this challenge as it pertains to fixed-wing applications, and it is somewhat premature to address this challenge as it pertains specifically to supersonic and hypersonic aircraft. The reference document for the SRW Project describes research related to this challenge, although issues related to fault detection, isolation, and reconfiguration of control are not explicitly pursued.

The IVHM Project is maturing health monitoring technologies related to structures, propulsion, and aircraft systems. It is developing assessment and prognostic capabilities for detecting airframe and propulsive structure faults using in situ sensors. The IVHM Project is developing high-temperature silicon carbide sensors for the propulsion system hot gas path. It is also studying the management and mitigation of icing- and lightning-related issues that relate to IVHM. Sensors, algorithms, and approaches are being developed for assessing landing gear, wings, propulsion systems, and power systems. The integration of these technologies, to form a "big picture" approach to vehicle health management, has been an elusive goal to date. The *Decadal Survey* is very clear on the need to integrate the health management functions of individual systems, and the IVHM Project is addressing this goal through the use of novel data mining techniques.

The External Hazard Detection and Classification element of the IIFD Project directly addresses fault tolerance. The Adaptive Information Management element of the IIFD Project supports research related to vehicle health to address vehicle situational awareness, along with the IVHM and IRAC Projects (see below), but the integration of these efforts is left to the researchers involved in the various efforts.

The IRAC Project is effectively and directly addressing fault tolerance due to structural anomalies and control actuation failures or degradation by integrating propulsion control and aerodynamic control in an adaptive control architecture. Fault tolerance after adaptation is enhanced by integrating a replanning function that considers the reduced functionality of the aircraft.

D6 Improved onboard weather systems and tools

SFW	SRW	Supersonics	Hypersonics	Airportal	Airspace	IVHM	IIFD	IRAC	Aging A/C
				■	■				

This R&T challenge has the following milestones:

- Develop robust and reliable data links for collecting information from onboard sensors.
- Develop processes and tools for integrating weather information from onboard sensors and data links to the ground and other aircraft.
- Demonstrate effectiveness in practical decision-support applications relating to weather, with varying levels of information quality and uncertainty.

Neither the NGATS ATM-Airportal Project nor the NGATS ATM-Airspace Project has planned research to achieve aspects of the above milestones related to in-flight applications.

The External Hazard Detection and Classification element of the IIFD Project is researching sensor suites for detection of adverse weather and obstacle avoidance while also investigating how best to display these data. This research is also addressing the feasibility of fusing alternative data sources that provide data on turbulence, icing conditions, and other weather-related phenomenology, and it is addressing display capabilities for runway incursion detection and obstacles in the air, such as thin wires. However, the IIFD Project does not include data link research.

D7 Advanced communication, navigation, and surveillance technology

SFW	SRW	Supersonics	Hypersonics	Airportal	Airspace	IVHM	IIFD	IRAC	Aging A/C
	■			■	■	■			

This R&T challenge has the following milestones:

- Simulate avionics on an individual aircraft to determine the capability of each avionics function (communication, navigation, guidance, control, and surveillance).
- Demonstrate (1) fault-tolerant degradation of communications, navigation, and surveillance capability (in terms of accuracy and availability of modes) and (2) processes needed to ensure that the individual aircraft can still transmit the needed aircraft state information and receive information and air traffic control commands with an extremely low probability of communication error.
- Evaluate different tracking and control algorithms with various faults that could occur in either the ATM system or airborne aircraft to determine whether the algorithms are able to detect the faults, identify them, and recover from them by reconfiguring the system in which the fault occurred as well as other systems to provide a satisfactory level of service.
- Document the feasibility of using space-based communications and surveillance as both a primary and backup means of ATM.
- Demonstrate modeling and real-time simulation using distributed control centers and different traffic levels, ranging from the current peak hourly load of about 6,000 airborne aircraft in the continental U.S. airspace to a predicted hourly load of 18,000 airborne aircraft, using current demand patterns. This effort is required to verify that the network of communication links, processing nodes in the network, and control algorithms provides the desired capacity while satisfying safety criteria.
- Demonstrate a means to provide seamless information flow between an aircraft's multiband antenna and the fiber-optic local area network that manages the information flow between aircraft systems and the radio channels.
- Demonstrate a robust IVHM system that detects permanent and transient onboard system faults and communicates system status to pilots and ground systems.
- For aircraft equipped with autothrottles, develop performance algorithms linked to aircraft dynamics to maintain the approved flight trajectory while minimizing fuel consumption. For aircraft that are not equipped with autothrottles, document the information required by the flight management system to generate speed commands to be displayed to pilots while minimizing pilot workload.
- Develop an air-ground communication protocol that (1) optimally allocates functions among pilots, avionics, air traffic controllers, and automated ground systems and (2) includes a means to alert ground systems and controllers that the data link or an onboard system has failed. This will require control algorithms that can handle multiple failures in terms of controlling the aircraft with the failures as well as adjacent traffic to minimize the impact on airspace capacity and efficiency.

Neither the NGATS ATM-Airportal Project, NGATS ATM-Airspace Project, nor IVHM Project has planned research to address the above milestones, and the SRW Project is not supporting research to address rotorcraft-specific aspects of this challenge.

D8 Human–machine integration

SFW	SRW	Supersonics	Hypersonics	Airportal	Airspace	IVHM	IIFD	IRAC	Aging A/C

This R&T challenge has the following milestones for human–machine integration *methods and tools:*

- Develop improved system engineering processes and tools for determining optimum roles of humans and automation in complex systems and demonstrate the benefits of this improved methodology in a trial application. This milestone should include provisions for dynamic human–machine task allocation and monitoring of human performance by machines (e.g., automated terrain avoidance).
- Conduct fundamental research on the causes of human error and on human contributions to safety and document design guidelines that will (1) help minimize the potential for design-induced error and (2) facilitate positive human intervention in the event of system failures. Transfer these guidelines to government program offices and industry.
- Develop constructive models of human performance and decision making and validate model predictions against objective performance data acquired in high-fidelity human-in-the-loop flight simulation experiments.
- Develop and demonstrate rapid prototyping tools that enable comparative evaluations of alternative automation schemes early in system development.
- Develop and validate a technique for integrating human reliability estimates into system safety and reliability analyses.

This R&T challenge also has the following milestones for human–machine integration technologies for vehicle applications:

- Develop and test enabling technologies for pilot workload management and reduced crew operations (e.g., improved human–machine integration for a flight management system) while keeping pilot awareness at the proper level.
- Develop display concepts for maintaining operator situational awareness while monitoring highly automated processes. Demonstrate the ability of operators to rapidly and accurately intervene in the event of system failures.
- Develop technologies and/or display concepts enabling effective fusion of information from multiple sources, including real-world and synthetic imagery (i.e., augmented reality). Demonstrate the effectiveness of these concepts in practical decision-support applications with varying levels of information quality and uncertainty (in terms of accuracy, timeliness, etc.).
- Develop and demonstrate technologies for machine vision (image-based object detection).
- Develop tools and metrics to compare effectiveness of machine and human operators in see-and-avoid tasks to improve machine performance.

The SRW Project is addressing some rotorcraft-specific aspects of human–machine interaction issues, but it does not include research to develop processes and tools to determine the appropriate balance between human operations and automation. Instead, existing research is focused on control methods to reduced pilot workload.

The Airportal Transition and Integration Management element of the NGATS ATM-Airportal Project substantially addresses the above milestones, particularly with regard to the allocation of roles and responsibilities between humans and machines as it applies to airport and terminal area systems and operations.

The NGATS ATM-Airspace Project includes separation assurance research to facilitate development of automated separation using a service provider approach. This research includes human situational awareness with automated systems through simulations, and it aligns with the above milestone to develop constructive models of human performance.

The Automation Monitoring and Failure Mitigation element and the Operator State Monitoring and Classification element of the IIFD Project include a very strong effort to investigate dynamic human–automation task allocation and to develop tools to measure the benefits of specific allocations and identify the causes of human error.

D9 Synthetic and enhanced vision systems

SFW	SRW	Supersonics	Hypersonics	Airportal	Airspace	IVHM	IIFD	IRAC	Aging A/C
							■		

This R&T challenge has the following milestones:

- Prepare an accurate and complete terrain and obstacles database and demonstrate real-time database monitoring and error correction.
- Develop procedures and rules for fusing image information from multiple imaging sensors as well as stored terrain data and traffic; identify common viewing parameters; and determine what role enhanced vision systems and synthetic vision systems should play in an integrated system.
- Demonstrate increased situational awareness and alerting to avoid air traffic, airport surface traffic, wires, and cables.
- Demonstrate displays that (1) eliminate image fusion artifacts that lead to misleading information and (2) present conformal information to pilots in a way that facilitates its transition to the outside world.
- Demonstrate tools for verifying database accuracy, fault tolerance, reliability, and overall system accuracy.

The External Hazard Detection and Classification element of the IIFD Project contributes to the development of synthetic and enhanced vision systems. Relevant research would provide the capability to maneuver in poor visibility on the ground and in the air (e.g., to support approach and landing operations on parallel or converging runways). This research also supports development of integrated displays of terrain and weather using onboard and external sensors and data sources.

D10 Safe operation of unmanned air vehicles in the national airspace

SFW	SRW	Supersonics	Hypersonics	Airportal	Airspace	IVHM	IIFD	IRAC	Aging A/C
				■	■			■	

This R&T challenge has the following milestones:

- Develop and demonstrate secure, reliable communications as well as procedures for interaction between UAVs and air traffic controllers.

- Design, develop, and demonstrate human interfaces for remote UAV operators under conditions extant in the air transportation system.
- Develop and test training programs for remote UAV operators.
- Develop and demonstrate sense-and-avoid technologies for UAVs.
- Demonstrate technologies for maintaining positive control of UAVs under adverse conditions.
- Develop and demonstrate automated contingency management for control of UAVs.

Neither the NGATS ATM-Airportal Project, NGATS ATM-Airspace Project, nor the IRAC Project has planned research to address the above milestones.

INTELLIGENT AND AUTONOMOUS SYSTEMS, OPERATIONS AND DECISION MAKING, HUMAN INTEGRATED SYSTEMS, AND NETWORKING AND COMMUNICATIONS

This section summarizes the committee's assessment of NASA research related to the top 10 R&T challenges involving intelligent and autonomous systems, operations and decision making, human integrated systems, networking, and communications (Area E) in the *Decadal Survey of Civil Aeronautics* (NRC, 2006).

E1 Methodologies, tools, and simulation and modeling capabilities to design and evaluate complex interactive systems

SFW	SRW	Supersonics	Hypersonics	Airportal	Airspace	IVHM	IIFD	IRAC	Aging A/C

This R&T challenge has the following milestones:

- Demonstrate methodologies and tools for the design, test, and certification of a flexible, robust, safe air transportation system that is readily adaptable to changing operational paradigms suited to new and different vehicles, including UAVs, very light jets, and spacecraft operating in civil airspace; communications, navigation, and surveillance capabilities; and optimization techniques.
- Demonstrate a flexible ATM model that incorporates the performance characteristics and limitations of the wide mix of present and future aircraft arriving, departing, and operating within airspace surrounding major hub airports. This model should be capable of analyzing the impacts of (1) aircraft mix and (2) operator and controller decision making and actions on system efficiency and capacity.
- Demonstrate the ability of an enhanced version of the model to assess the impact of regional weather phenomena, such as convective activity, snow, and high winds.
- Demonstrate the capability to test and certify nondeterministic systems.
- Demonstrate the ability of an enhanced version of the ATM model to assess impacts of aircraft mix and operator and controller decision making.

The NGATS ATM-Airportal Project is a contributor to and user of simulation facilities and laboratories operated by NASA. Plans for the Coordinated Approach/Departure Operations Management element of the project include research to produce a suite of concepts and technologies that can be tailored to specific airports. In addition, plans for the Safe and Efficient Surface Operations element include research to develop and validate a surface 4-D trajectory model. Simulation and modeling underpin most of the work by this project.

The Performance Based Services element and the System-Level Design, Analysis, and Simulation Tool element of the NGATS ATM-Airspace Project substantially address most of the above milestones. Research in several other Airspace elements also supports this challenge, although none of the planned work specifically addresses unmanned air vehicles or spacecraft operating in civil airspace. In some cases, resource limitations, including the availability of sufficient in-house expertise, will make it difficult to achieve milestones in a timely fashion.

The IRAC and IVHM Projects have plans to address the milestone regarding the testing of nondeterministic systems. The IRAC Project plans to conduct laboratory and flight tests of nondeterministic systems at Dryden Flight Research Center, although not until late in the current 10-year plan. Related work by the IVHM Project will piggyback on the IRAC flight tests at Dryden.

E2 New concepts and methods of separating, spacing, and sequencing aircraft

SFW	SRW	Supersonics	Hypersonics	Airportal	Airspace	IVHM	IIFD	IRAC	Aging A/C

This R&T challenge has the following milestones:

- Demonstrate high-efficiency airspace and airway structures that can be effectively managed and understood.
- Design and evaluate separation, spacing, and sequencing procedures for UAVs operating in civilian airspace and assess their impact on commercial aircraft capacity and safety.
- Extend models and simulation tools to enable accurate evaluation of emerging technologies (e.g., the Automatic Dependent Surveillance-Broadcast system) in all weather conditions and during all phases of flight.
- Complete an in-depth examination of the ability of concepts such as runway-independent aircraft and UAV formations or swarms to safely increase capacity and accommodate nontraditional aircraft operations.
- Demonstrate advanced, autonomous collision avoidance technologies and protocols.

Plans for the Safe and Efficient Surface Operations element of the NGATS ATM-Airportal Project address surface scheduling and taxi routes. Plans for the Coordinated Approach/Departure Operations Management element include research to study vortex avoidance, and runway balancing, assignments, and self-spacing may be addressed in the future. Plans for the Airportal Transition and Integration Management element also foresee including research on metroplex operations (i.e., areas with two or more airports in close proximity) in the future.

The Separation Assurance and Super Density Operations elements of the NGATS ATM-Airspace Project address each of the above milestones except for those related to unmanned air vehicles operating in civil airspace. Ongoing work related to this challenge is likely to advance the state of the art, but available resources will make it difficult to achieve milestones in a timely fashion. None of the planned work specifically addresses unmanned air vehicles.

E3 Appropriate roles of humans and automated systems for separation assurance, including the feasibility and merits of highly automated separation assurance systems

SFW	SRW	Supersonics	Hypersonics	Airportal	Airspace	IVHM	IIFD	IRAC	Aging A/C

This R&T challenge has the following milestones:

- Complete basic research necessary to determine the most appropriate separation assurance roles for humans and automation, for ground-centered and aircraft-centered designs.
- Complete the development of the NASA Ames Advanced Airspace Concept, an automated ground-based separation assurance system, for the en route domain.
- Determine how humans interact with the Advanced Airspace Concept and other automation designs.
- Determine how the Advanced Airspace Concept and other designs respond to air and/or ground automation failures, or when the flight crew fails to respond to automated directives.
- Develop an adaptation of the Advanced Airspace Concept or other designs for UAVs, and determine its performance.
- Determine through analysis and simulation the safety of the Advanced Airspace Concept and other designs.

The Airportal Transition and Integration Management element of the NGATS ATM-Airportal Project includes a substantial, well-conceived effort to address human–system integration as it applies to the complex airportal domain. However, plans for the Coordinated Approach/Departure Operations Management element have pushed the investigation of high-density operations and separation assurance in airport and terminal areas well into the future.

The Separation Assurance and Super Density Operations elements of the NGATS ATM-Airspace Project include research on the roles of humans and automated systems, but this is not a key focus of these elements. Ongoing work related to this challenge would advance the state of the art, but available resources will make it difficult to achieve milestones in a timely fashion. None of the planned work specifically addresses unmanned air vehicles.

E4 Affordable new sensors, system technologies, and procedures to improve the prediction and measurement of wake turbulence

SFW	SRW	Supersonics	Hypersonics	Airportal	Airspace	IVHM	IIFD	IRAC	Aging A/C
				■	■				

This R&T challenge has the following milestones:

- Demonstrate new sensors, including a scientific, coherent lidar capable of accurate wake velocity strength measurements.
- Conduct phenomenological studies of wake behavior supported by field experiments using ground-based sensor(s) that measure wake decay and atmospheric conditions at altitudes up to 8,000 feet above the ground.

- Determine aircraft upset risks from wake vortices encounters, taking advantage of existing models and enhancing them where needed with field data.
- Demonstrate procedures, monitoring equipment, and other systems to safely reduce wake separation.
- Demonstrate an airborne means to sense and quantify the intensity of hazardous wakes en route in time for aircraft to evade them.

Plans for the Coordinated Approach/Departure Operations Management element of the NGATS ATM-Airportal Project include exploration of "wake aware" procedures. However, the Airportal Project views wake vortex issues as one of many constraints on runway, airport, and terminal area capacity, and the scope of this research has been reduced in recent years. NASA does not plan to develop or demonstrate new sensors or sensor technologies. NASA plans to rely on the National Oceanic and Atmospheric Administration and the FAA to take the lead in research related to the determination and characterization of weather and hazards such as wake vortices.[4] The NGATS ATM-Airportal Project is supporting foundational research (e.g., mathematical and statistical characterization) to support some of the above milestones, but these activities are not planning to proceed to the point of achieving the above milestones. The resulting work would be handed off to the National Oceanic and Atmospheric Administration and the FAA.

The NGATS ATM-Airspace Project has no planned research to address the above milestones.

E5 Interfaces that ensure effective information sharing and coordination among ground-based and airborne human and machine agents

SFW	SRW	Supersonics	Hypersonics	Airportal	Airspace	IVHM	IIFD	IRAC	Aging A/C
				■					

This R&T challenge has the following milestones:

- Document improved understanding of human cognitive control, judgment, and decision making in a variety of contexts and under a variety of stressors.
- Document improved understanding of organizational dynamics and business concerns associated with information sharing.

Plans for the Safe and Efficient Surface Operations element of the NGATS ATM-Airportal Project include research on multiagent decision making and trajectory conformance for surface operations. Plans for the Coordinated Approach/Departure Operations Management element include the study of 4-D trajectory conformance and investigation of separation assurance in super density situations. Plans for the Airportal Transition and Integration Management element also include investigation of human–system integration issues associated with this domain.

Plans for the Separation Assurance and Super Density Operations elements of the NGATS ATM-Airspace Project include substantial research on humans-in-the-loop information sharing. However, research by the Airportal and Airspace projects will not address the milestone on organizational dynamics and business concerns.

[4]Another concurrent National Research Council study is focused exclusively on NASA's wake vortex research. The results of that study were not available in time to factor them into this report.

E6 Vulnerability analysis as an integral element in the architecture design and simulations of the air transportation system

SFW	SRW	Supersonics	Hypersonics	Airportal	Airspace	IVHM	IIFD	IRAC	Aging A/C

This R&T challenge has the following milestones:

- Complete end-to-end vulnerability analysis of system architecture and signal flow.
- Demonstrate the ability of a more capable model to simulate critical element disruptions as defined by vulnerability analyses.
- Document safety and capacity impacts using modified system simulations.
- Develop changes in system architecture and operational procedures and demonstrate that they can mitigate the effects of specific system disruptions.

The NGATS ATM-Airportal Project is monitoring relevant work being done by the NGATS ATM-Airspace Project, but it has no planned research to address the above milestones.

The NGATS ATM-Airspace Project does not contemplate development of a complete end-to-end vulnerability analysis or changes in system architecture and operational procedures. Plans for the Separation Assurance and Super Density Operations elements include modeling work that would factor in system disruptions. This research would advance the state of the art, but available resources will make it difficult to achieve milestones in a timely fashion.

E7 Adaptive ATM techniques to minimize the impact of weather by taking better advantage of improved probabilistic forecasts

SFW	SRW	Supersonics	Hypersonics	Airportal	Airspace	IVHM	IIFD	IRAC	Aging A/C

This R&T challenge has the following milestones:

- Identify potential reductions in weather-induced delays.
- Demonstrate use of automated weather forecasts in making traffic flow decisions.
- Quantify the benefit of using automated weather forecasts in making traffic flow decisions.
- Determine where this capability is cost-beneficial.

The NGATS ATM-Airportal Project is very attuned to the need for adaptive ATM techniques. Plans for the Safe and Efficient Surface Operations element include research to develop surface optimization schemes that would continually adapt to produce the optimal result. Plans for the Coordinated Approach/Departure Operations Management element include research to study the effect of weather on wake vortex behavior and, in the future, to develop better approaches for runway balancing and reconfiguration and for equivalent visual operations (which would allow aircraft to operate regardless of visibility conditions or the ability to make direct visual observations). The Airportal Transition and Integration Management element also foresees including research on metroplex operations in the future. The NGATS ATM-Airportal Project is not performing cost-benefit analyses.

The Traffic Flow Management element of the NGATS ATM-Airspace Project seeks to develop models to forecast demand and capacity of the National Airspace System that respond effectively to weather uncertainties. Assuming success, these models could help to identify potential reductions in weather-induced delays. It does not appear, however, that the remainder of the challenge milestones are planned to be addressed.

E8a Transparent and collaborative decision support systems

SFW	SRW	Supersonics	Hypersonics	Airportal	Airspace	IVHM	IIFD	IRAC	Aging A/C

This R&T challenge has the following milestones:

- Identify the type of information to be shared between human operators and automated decision-support systems and the most appropriate form of information representation and exchange.
- Develop, demonstrate, evaluate, and iteratively refine candidate designs in collaboration with operators.

Plans for the Safe and Efficient Surface Operations element of the NGATS ATM-Airportal Project include substantial work to develop concepts for decision-support methodologies and tools applicable to surface operations.

Plans for the Traffic Flow Management element of the NGATS ATM-Airspace Project include research to address both of the above milestones. This research would likely advance the state of the art and produce timely results.

E8b Using operational and maintenance data to assess leading indicators of safety

SFW	SRW	Supersonics	Hypersonics	Airportal	Airspace	IVHM	IIFD	IRAC	Aging A/C

This R&T challenge has the following milestones:

- Produce a common taxonomy for all safety information acceptable to all stakeholders.
- Demonstrate methodologies to discover and analyze anomalous system, components, and human behavior in nominal and off-nominal conditions.
- Demonstrate methods to integrate system models into analytical processes.
- Demonstrate advanced, affordable methods to analyze anecdotal written reports of safety problems and cross-reference them to operational data from aircraft, ATM, and weather systems.
- Demonstrate methods to cross-reference operational data to certification and training simulator data to determine if aircraft are performing as designers intended and if pilots and controllers are performing as trained.

The IVHM Project plans to support extensive research in data mining and analysis of aircraft maintenance records as well as operational data. This information would be fused with sensor data from the vehicle airframe; the IVHM Project is working with the IIFD Project to assess integration of this work into the cockpit.

Plans for each of the eight challenge problems being addressed by the Aging Aircraft and Durability Project include research to help create a database that would provide better indicators of safety. However, the challenge problems do not directly discuss maintenance, and it seems unlikely that planned research will have a significant and timely impact without additional resources and the involvement of personnel experienced with airline maintenance and operations.

E8c Interfaces and procedures that support human operators in effective task and attention management

SFW	SRW	Supersonics	Hypersonics	Airportal	Airspace	IVHM	IIFD	IRAC	Aging A/C

This R&T challenge has the following milestones:

- Complete basic research to document how operators absorb information, process information, and prioritize tasks.
- Demonstrate tools to efficiently evaluate operational data and reports of nominal and off-nominal decision making by operators.
- Demonstrate and evaluate candidate designs and procedures in support of preattentive reference, time-sharing among different tasks, and task switching.[5]

The Airportal Transition and Integration Management element of the NGATS ATM-Airportal Project includes an in-depth and comprehensive research effort to address the role of humans.

The NGATS ATM-Airspace Project has no planned research to address the above milestones except as a by-product of the research by the Separation Assurance and Super Density Operations elements.

SPACE AND NON-CIVIL AERONAUTICS RESEARCH

In addition to the civil aeronautics R&T challenges detailed above, the committee for this study identified two high-priority requirements for NASA aeronautics research based on NASA's own requirements for aeronautics research (including robotic and human space exploration) and the needs of other federal government departments and agencies for non-civil aeronautics research.

S-1 Entry, descent, and landing on Mars (e.g., high-Mach-number parachutes) and Earth (ablative materials)

SFW	SRW	Supersonics	Hypersonics	Airportal	Airspace	IVHM	IIFD	IRAC	Aging A/C

The Supersonics Project is supporting computational and experimental investigations of aerothermodynamics issues associated with entry, descent, and landing, as well as studies of supersonic aerodynamic decelerators.

[5]Preattentive reference is supported by presenting partial information about a potentially interrupting task or event to help the operator decide whether a shift in attention is warranted. The information needs to be presented in such a way that it is quickly noticed and easily processed and understood without requiring an interruption of the ongoing task or line of reasoning (Woods, 1995). Operational systems that provide preattentive reference reduce the risk of task switching errors and improve operator efficiency and performance.

The Hypersonics and Supersonics Projects both include research that supports robotic and human space exploration. The study of High Mass Mars Entry Systems constitutes about 25 percent of the Hypersonics Project. This effort is conducting fundamental research on issues related to landing high-mass payloads on Mars. Issues of interest include increased levels of turbulent and radiative heating caused by a larger entry capsule and the need for increased precision in landing accuracy. Relevant research is incorporated in the Aerodynamics, Aerothermodynamics, and Plasmadynamics; Materials and Structures; and Guidance, Navigation and Control elements of the Hypersonics Project, and it includes development of CFD analysis tools, experimental investigations, aerothermodynamics modeling, materials for thermal protection, and trajectory analysis.

S-2 Core competencies (facilities and staff) for space (e.g., for access to space, entry, descent, and landing, and analysis of space shuttle anomalies) and for DoD (hypersonic vehicles)

SFW	SRW	Supersonics	Hypersonics	Airportal	Airspace	IVHM	IIFD	IRAC	Aging A/C

NASA's Exploration Systems Mission Directorate, Science Mission Directorate, and Space Operations Mission Directorate all support important space activities. As noted above, in some cases NASA space programs rely on the NASA Aeronautics Program to sponsor research tasks of direct benefit to future space programs. In other cases, research, development, and operational elements of the space program rely on NASA aeronautics to support space research tasks by providing access to aeronautical staff and facilities.[6]

In addition, elements of the Department of Defense often rely on NASA Aeronautics to provide support, in the terms of key staff expertise and/or facilities, to support DoD research and development tasks. (It is much rarer for the DoD to ask NASA to conduct a research project on behalf of the DoD.) The chief scientist of the Air Force and the DARPA director report that they are satisfied with the cooperative support that NASA currently provides.

Interactions with DARPA have focused on the four projects that make up the Fundamental Aeronautics Program. Ongoing and recently completed collaborations include the following:

- Subsonic Fixed Wing Project
  —DARPA Morphing Wing
- Subsonic Rotary Wing Project
  —Helicopter Quieting Program
  —SMART Rotor Program
  —Heliplane
  —Acoustic flight tests at Eglin Air Force Base (co-sponsored with the Air Force Army)
  —Helicopter Brownout (co-sponsored with Air Force)
- Supersonics Project
  —Oblique Flying Wing Program
- Hypersonics Project
  —X-51 (1, 2, 3, 4) (co-sponsored with the Air Force)
  —Falcon
  —HyFly (co-sponsored with the Navy)

[6]See Chapter 3 for a detailed discussion of aeronautical facilities.

With many of the above projects, NASA subject-matter experts assist DARPA in the formulation of programs, review of proposals, and periodic program reviews. In addition, NASA subject-matter experts often use NASA test facilities (e.g., wind tunnels, flight-test operations, and propulsion test facilities) to support of DARPA programs. In some cases, NASA research in support of DARPA projects is funded by DARPA, and in other cases NASA contributes its own resources to support research of mutual benefit.

ASSESSMENT OF NASA'S RESPONSE TO RECOMMENDATIONS IN THE *DECADAL SURVEY OF CIVIL AERONAUTICS*

The *Decadal Survey of Civil Aeronautics* made eight recommendations (NRC, 2006, p. 3). The committee's assessment of NASA's response to these recommendations is summarized below.

Recommendation 1 from the *Decadal Survey*

> NASA should use the 51 Challenges listed in Table ES-1 as the foundation for the future of NASA's civil aeronautics research program during the next decade.[7]

Assessment of NASA Response. The content of the *Decadal Survey of Civil Aeronautics* seems not to have been a significant factor in the selection of the research portfolio being pursued by many of the ARMD's research projects. The basic planning documents for most of NASA's research projects were prepared before the *Decadal Survey* was published, and the NASA research portfolio, as a whole, does not seem to have changed course in response to the *Decadal Survey*. In any case, as detailed above, NASA is doing a mixed job in responding to the R&T challenges overall and in each R&T challenge area. As discussed in Chapter 4, addressing all 51 highest-priority R&T challenges from the *Decadal Survey* in a thorough and comprehensive manner would require a substantial increase in the NASA ARMD funding levels. Absent such an increase in funding and/or a substantial reduction in the constraints that NASA faces in conducting aeronautics research, NASA, in consultation with the aeronautics research community and others as appropriate, should redefine the scope and priorities of the aeronautics research program, even if all 51 of the highest-priority R&T challenges from the *Decadal Survey of Civil Aeronautics* are not addressed simultaneously.

Recommendation 2 from the *Decadal Survey*

> The U.S. government should place a high priority on establishing a stable aeronautics R&T plan, with the expectation that the plan will receive sustained funding for a decade or more, as necessary, for activities that are demonstrating satisfactory progress.

Assessment of NASA Response. NASA leadership issued a new vision for aeronautics research in early 2006, in the context of an agency vision that remains focused on space. For the next 2 years, NASA was consistent in advocating a research program that is stable, year to year, in carrying out the vision. However, changes in direction often occur with a change in leadership, and the associate administrator for ARMD who oversaw the creation and implementation of this vision left NASA in February 2008. It remains to be seen if the appointment of a permanent replacement will result in another change in NASA's vision and/or priorities for aeronautics research.

---

[7]These 51 challenges are listed in Table 1-1 in Chapter 1 of the present report.

Recommendation 3 from the *Decadal Survey*

NASA should use five Common Themes to make the most efficient use of civil aeronautics R&T resources:

- Physics-based analysis tools
- Multidisciplinary design tools
- Advanced configurations
- Intelligent and adaptive systems
- Complex interactive systems

Assessment of NASA Response. NASA has addressed the recommended common themes with different levels of success. NASA is doing well, in most cases, with physics-based analysis tools and multidisciplinary design tools. NASA is doing good work on advanced configurations for subsonic fixed-wing aircraft and, to a lesser extent, on hypersonic aircraft. It is not doing noteworthy work on advanced configurations for supersonic aircraft or rotorcraft. A stronger focus on systems and systems integration is necessary to strengthen research related to advanced configurations, intelligent and adaptive systems, and complex interactive systems.

Recommendation 4 from the *Decadal Survey*

NASA should support fundamental research to create the foundations for practical certification standards for new technologies.

Assessment of NASA Response. In many cases, NASA is supporting fundamental research that could ultimately be used to create foundations for practical certification standards for future technologies, but it is not conducting research specifically focused on certification issues. In addition, procedures for transferring new technologies are not apparent. However, Section 905 of H.R. 2881, the FAA Reauthorization Act of 2007, would direct the "FAA, in consultation with other agencies as appropriate" to "establish a research program on methods to improve both confidence in and the timeliness of certification of new technologies." The same bill would direct the FAA to prepare a research plan for this activity and to have the plan reviewed by the National Research Council. As of October 19, 2007, the bill had passed the House of Representatives and was awaiting action by the Senate.

Recommendation 5 from the *Decadal Survey*

The U.S. government should align organizational responsibilities as well as develop and implement techniques to improve change management for federal agencies and to assure a safe and cost-effective transition to the air transportation system of the future.

Assessment of NASA Response. Recommendation 5 is not directed at NASA, and it is beyond the scope of NASA's authority. This committee is not aware of any action to implement this recommendation (beyond the ongoing work of the NextGen Joint Planning and Development Office).

Recommendation 6 from the *Decadal Survey*

NASA should ensure that its civil aeronautics R&T plan features the substantive involvement of universities and industry, including a more balanced allocation of funding between in-house and external organizations than currently exists.

Assessment of NASA Response. The *Decadal Survey of Civil Aeronautics* reported that ARMD had plans to allocate only 7 percent of its budget for research by outside organizations. NASA is gradually increasing the involvement of universities and industry in ARMD research projects using NRAs. This percentage varies among the projects from 10 percent to more than 40 percent.[8]

Recommendation 7 from the *Decadal Survey*

NASA should consult with non-NASA researchers to identify the most effective facilities and tools applicable to key aeronautics R&T projects and should facilitate collaborative research to ensure that each project has access to the most appropriate research capabilities, including test facilities; computational models and facilities; and intellectual capital, available from NASA, the FAA, the Department of Defense, and other interested research organizations in government, industry, and academia.

Assessment of NASA Response. NASA is collaborating effectively with the Department of Defense in facility management. NASA is also improving collaboration with other research organizations in some areas.

Recommendation 8 from the *Decadal Survey*

The U.S. government should conduct a high-level review of organizational options for ensuring U.S. leadership in civil aeronautics.

Assessment of NASA Response. Recommendation 8 is not directed at NASA, and it is beyond the scope of NASA's authority. However, Section 604 of S.1300, the Aviation Investment and Modernization Act of 2007, would direct the President to establish the Advisory Committee on the Future of Aeronautics. The committee would consist of 7 members selected from a list of 15 candidates proposed by the National Academy of Sciences. The committee would "examine the best governmental and organizational structures for the conduct of civil aeronautics research and development, including options and recommendations for consolidating such research to ensure continued United States leadership in civil aeronautics. The Committee shall consider transferring responsibility for civil aeronautics research and development from NASA to other existing departments or agencies of the Federal government or to a non-governmental organization such as academic consortia or not-for-profit organizations." As of October 30, 2007, this bill was still under consideration by the Senate.

[8]In Chapter 3, see the section entitled "Aeronautics Workforce Issues" for more information related to the involvement of external organizations in NASA's research.

REFERENCES

NRC (National Research Council). 2006. Decadal Survey of Civil Aeronautics: Foundation for the Future. Washington, D.C.: The National Academies Press. Available online at <http://www.nap.edu/catalog.php?record_id=11664>.

NASA (National Aeronautics and Space Administration). 2006. Aviation Safety Program, Aircraft Aging & Durability Project Technical Plan Summary. Washington, D.C.: NASA Headquarters, Aeronautics Research Mission Directorate. Available online at <www.aeronautics.nasa.gov/nra_pdf/aad_technical_plan_c1.pdf>.

OAG (OAG Worldwide Limited). 2007. Trends in fleet models and aircraft age. London, England: OAG Worldwide Limited. Available online at <http://www.oag.com/oagcorporate/press_releases_overview.html>.

Woods, D.D. 1995. The alarm problem and directed attention in dynamic fault management. Ergonomics 38(11): 2371-2393.

Wlezien, R., and L. Veitch. 2002. Quiet Supersonic Platform. AIAA Paper 2002-0143. January 2002. Washington, D.C.: American Institute of Aeronautics and Astronautics.

3

Workforce and Facilities

This chapter addresses the adequacy of NASA and national workforce and facilities, particularly with regard to achieving the goals of the NASA aeronautics research program and meeting the research and technology (R&T) challenges described in Chapter 2.

AERONAUTICS WORKFORCE ISSUES

National Aeronautics Workforce

National employment data give a mixed picture of aerospace engineering employment and trends—with no consistent trends in terms of employment numbers or salaries. From 1996 to 2004, aerospace engineering employment increased 43 percent, and this was well above the trend for total engineering employment, which increased only 5 percent over the same time period (see Table 3-1a). However, aerospace engineering employment decreased from 2002 to 2005 (see Tables 3-1b and 3-2a) before rebounding again in 2006 (see Table 3-2b).[1] Aerospace engineering salaries were almost flat from 2004 to 2005, as employment fell by 16 percent (Table 3-2a), but then increased by 11 percent from 2005 to 2006, as employment grew by 10 percent (Table 3-2b). In addition, in 2006, median salaries for aerospace engineers were higher than for any other professional occupation tracked by annual reports of median wages prepared by the Bureau of Labor Statistics (BLS), except for lawyers, judges, physicians and surgeons, and pharmacists (BLS, 2007).

In 2006, based on detailed employment data from 2004, the BLS concluded the following regarding the future of aerospace engineering employment: "Aerospace engineers are expected to have slower-than-

[1]All the data in Tables 3-1 and 3-2 are taken from documents produced by the Department of Labor's Bureau of Labor Statistics (BLS). (Sources are listed below each table.) The data for Table 3-1 are taken from different annual editions of the same document, and the data for Table 3-2 are taken from different annual editions of a different document. Thus, the data within each table are internally consistent, but there are some inconsistencies between the data in Table 3-1 (covering 1996 to 2004) and the data in Table 3-2 (covering 2004 to 2006). Therefore, data from the different tables should not be combined to derive overall trends from 1996 to 2006.

TABLE 3-1a Changes in Engineering Employment Between 1996 and 2004

	Number of Engineers			
	1996	2004	Change	Change (%)
Total, all engineers	**1,382,000**	**1,450,100**	**68,100**	**5**
Mining	3,000	5,200	2,200	73
Industrial	115,000	177,000	62,000	54
Aerospace	53,000	76,000	23,000	43
Petroleum	13,000	16,000	3,000	23
Nuclear	14,000	17,000	3,000	21
Civil	196,000	237,000	41,000	21
Materials	18,000	21,000	3,000	17
Electrical, Electronics, and Computer	367,000	376,000	9,000	2
Mechanical	228,000	226,000	–2,000	–1
Chemical	49,000	31,000	–18,000	–37
All other engineers	326,000	267,900	–58,100	–18

NOTE: Growth in aerospace engineering employment ranked 3 out of 10 from 1996 to 2004.

TABLE 3-1b Changes in Engineering Employment Between 2002 and 2004

	Number of Engineers			
	2002	2004	Change	Change (%)
Total, all engineers	**1,478,000**	**1,449,000**	**–29,000**	**–2.0**
Marine engineers and naval architects	5,000	6,800	1,800	36.0
Biomedical engineers	8,000	9,700	1,700	21.3
Petroleum engineers	14,000	16,000	2,000	14.3
Agricultural engineers	3,000	3,400	400	13.3
Nuclear engineers	16,000	17,000	1,000	6.3
Industrial engineers	194,000	204,000	10,000	5.2
Mechanical engineers	215,000	226,000	11,000	5.1
Environmental engineers	47,000	49,000	2,000	4.3
Computer hardware engineers	74,000	77,000	3,000	4.1
Mining and geological engineers	5,000	5,200	200	4.0
Civil engineers	228,000	237,000	9,000	3.9
Electrical and electronics engineers	292,000	299,000	7,000	2.4
Aerospace engineers	78,000	76,000	–2,000	–2.6
Chemical engineers	33,000	31,000	–2,000	–6.1
Materials engineers	24,000	21,000	–3,000	–12.5
All other engineers	243,000	172,000	–71,000	–29.2

NOTE: Growth in aerospace engineering employment ranked 13 out of 15 from 2002 to 2004.

SOURCES:

1996 data: George T. Silvestri, Office of Employment, Projections, Bureau of Labor Statistics, Employment outlook: 1996-2006, Occupational employment projections to 2006, Monthly Labor Review, November 1997, pp. 58-82 (Table 1, pp. 59-60). Available online at <www.bls.gov/emp/empbib05.htm>.

2002 data: Daniel Hecker, Occupational employment projections to 2012, Monthly Labor Review, February 2004, pp. 80-105 (Table 2, pp. 82ff). Available online at <www.bls.gov/opub/mlr/2004/02/art5full.pdf>.

2004 data: BLS, 2006, Bureau of Labor Statistics, U.S. Department of Labor, Occupational Outlook Handbook, 2006–07 Edition, Engineers.

TABLE 3-2a Changes in Annual Average of Employment Numbers and Weekly Earnings Between 2004 and 2005

	Number of Workers		Change	Full-time Median Weekly Earnings ($)		Change
	2004	2005	(%)	2004	2005	(%)
Architecture and engineering occupations	**2,500,000**	**2,509,000**	**0.4**	**1,098**	**1,105**	**0.6**
Chemists and materials scientists	133,000	109,000	−18.0	1,048	1,128	7.6
Mechanical engineers	292,000	306,000	4.8	1,187	1,262	6.3
Computer hardware engineers	86,000	72,000	−16.3	1,328	1,405	5.8
Electrical and electronics engineers	311,000	330,000	6.1	1,277	1,350	5.7
Aerospace engineers	105,000	88,000	−16.2	1,347	1,362	1.1
Industrial engineers, including health and safety	178,000	185,000	3.9	1,152	1,161	0.8
Architects, except naval	142,000	176,000	23.9	1,141	1,146	0.4
Civil engineers	264,000	277,000	4.9	1,135	1,138	0.3

NOTE: Aerospace engineers' change in employment, 2004-2005: rank 6 of 8 (shrinking); change in salary, 2004-2005: rank 5 of 8 (close to zero).

TABLE 3-2b Changes in Annual Average of Employment Numbers and Weekly Earnings Between 2005 and 2006

	Number of Workers		Change	Full-time Median Weekly Earnings ($)		Change
	2005	2006	(%)	2005	2006	(%)
Architecture and engineering occupations	**2,509,000**	**2,568,000**	**2.4**	**1,105**	**1,155**	**4.5**
Aerospace engineers	88,000	97,000	10.2	1,362	1,508	10.7
Civil engineers	277,000	276,000	−0.4	1,138	1,251	9.9
Electrical and electronics engineers	330,000	362,000	9.7	1,350	1,386	2.7
Industrial engineers, including health and safety	185,000	162,000	−12.4	1,161	1,175	1.2
Chemists and materials scientists	109,000	121,000	11.0	1,128	1,131	0.3
Mechanical engineers	306,000	316,000	3.3	1,262	1,253	−0.7
Architects, except naval	176,000	161,000	−8.5	1,146	1,112	−3.0
Computer hardware engineers	72,000	76,000	5.6	1,405	1,292	−8.0

NOTE: Aerospace engineers' change in employment, 2005-2006: rank 2 of 8 (growing); change in salary, 2005-2006: rank 1 of 8 (growing).

SOURCES:

2004 data: BLS, 2005, Median weekly earnings of full-time wage and salary workers by detailed occupation and sex, 2004. Available online at <www3.ccps.virginia.edu/career_prospects/Statistics/National/EmpGender2004.pdf>.

2005 data: BLS, 2006, Median usual weekly earnings of full-time wage and salary workers by detailed occupation and sex, 2005 annual averages. Available online at <www.bls.gov/cps/wlf-table18-2006.pdf>.

2006 data: BLS, 2007, Median weekly earnings of full-time wage and salary workers by detailed occupation and sex, 2006. Available online at <www.bls.gov/cps/cpsaat39.pdf>.

average growth in employment over the projection period. Although increases in the number and scope of military aerospace projects likely will generate new jobs, increased efficiency will limit the number of new jobs in the design and production of commercial aircraft. Even with slow growth, the employment outlook for aerospace engineers through 2014 appears favorable: the number of degrees granted in aerospace engineering declined for many years because of a perceived lack of opportunities in this field, and, although this trend is reversing, new graduates continue to be needed to replace aerospace engineers who retire or leave the occupation for other reasons" (BLS, 2006, p. 11).

In other words, the national aerospace industry will have an ongoing demand for new engineers that will probably be able to absorb all the new aerospace engineering graduates that U.S. universities produce, even as outsourcing by U.S. industry to foreign locations continues. However, much of the demand for new staff by U.S. industry will result from the retirement of existing workers, and the size of the aerospace workforce is not expected to grow very much. This cautious prediction seems consistent with mixed signals coming out of historical employment data for aerospace engineering and the uncertain cause of the sometimes large swings in employment that the data report. The situation specifically with regard to aeronautics and aeronautical engineers is even more uncertain, given that all of the data presented above and in the tables are for aerospace engineering as a whole, with no breakdown among various applications such as civil aeronautics, military aeronautics, and satellite and space applications.

NASA Aeronautics Workforce

Several recent reports have been issued which address the NASA workforce issue, with particular emphasis on the President's Vision for Space Exploration (White House, 2004), but none has explicitly addressed the NASA aeronautics program and its requirements. These reports include *NASA: Human Capital Flexibilities for the 21st Century Workforce,* issued by the National Academy of Public Administration (NAPA, 2005); *Building a Better NASA Workforce: Meeting the Workforce Needs for the National Vision for Space Exploration,* issued by the National Research Council (NRC, 2007); and the *National Aeronautics and Space Administration Workforce Strategy,* issued by NASA (2006b). These reports discuss in detail many key workforce issues, such as workforce demographics; trends in university enrollment and degrees awarded; the types of data necessary to support the creation of viable workforce strategies; NASA workforce competency requirements; recruitment, retention, training, and retraining of NASA workers; and the role of academia. They also contain the following recommendations:

- Recommendations to NASA from the National Academy of Public Administration (NAPA, 2005):
  —Integrate NASA's Leadership Development Program evaluation and benchmarking activities.
  —Accelerate benchmarking with exemplary private sector organizations and leading university-based development programs.
  —Respond effectively to Section 201 of the Federal Workforce Flexibility Act 2004 (Public Law No. 108-411).[2]

- Recommendations to NASA from the National Research Council (NRC, 2007):
  —Collect detailed data on NASA workforce requirements.
  —Hire and retain younger workers within NASA.

[2]Section 201 directs federal agencies to regularly evaluate and modify training programs and plans to promote a strategic approach to the integration of training programs into overall agency missions.

—Ensure a coordinated national strategy for aerospace workforce development among relevant institutions.
—Provide hands-on training opportunities for NASA workers.
—Support university programs and provide hands-on opportunities at the college level.
—Support involvement in suborbital programs and nontraditional approaches in developing skills.

- Recommendations from NASA (2006b):
  —Build and maintain 10 strong healthy centers.
  —Integrate and emphasize retraining of the current workforce.
  —Fully develop and maintain a structural workforce training process.
  —Use term and temporary hiring authorities.
  —Maintain a high-quality workplace.
  —Apply recruitment flexibilities.
  —Offer financial incentives.
  —Manage surplus employees.
  —Institute hiring controls.
  —Foster knowledge management.

The above recommendations present a long-range strategic approach to NASA's workforce issues as a whole, with considerable emphasis on space exploration. Even so, these recommendations also seem to be generally applicable to the aeronautics workforce, and most of the recommendations could be implemented, at least in part, within the existing budget for NASA and its Aeronautics Research Mission Directorate (ARMD). However, nationally and at NASA, the issues faced by the aeronautics workforce differ in some key respects from the issues faced by the aerospace workforce as a whole. For example, a workforce plan that excludes the possibility of consolidating and reducing the number of centers is probably best justified not in terms of workforce efficiency and effectiveness, but in terms of the political reality that Congress seems highly unlikely to approve such a reduction.

The principal investigators (PIs) for each of the 10 ARMD research projects were individually interviewed by the committee and asked to respond to the following questions:

- Do you foresee, in your area of responsibility, any lack of skills, etc., necessary to perform the research called out in your project plan?
- What are your staffing/research workforce plans for the future in order to effectively perform the research called out in your project plan?
- Is there any specific ARMD workforce-related strategy/implementation plan in place and operational?
- What are the specific impacts/constraints, etc., if any, emanating from the Office of Management and Budget, NASA Headquarters, and/or NASA Center Directors regarding staffing?

Not all PIs responded to all of the preceding questions, particularly the second one. Also, the answers varied somewhat among specific PIs and projects. However, in general the responses were similar enough that general trends and findings could be established.

Workforce issues are being addressed project by project. Each PI manages assigned resources as best fits the requirements of his or her project plan. Authorized staffing profiles are negotiated between the NASA Headquarters program managers and the center-resident PIs. In addition, the center directors have final approval over the hiring of any permanent civil service employees at their centers. A better approach would be for ARMD to develop a comprehensive strategic plan that addresses the needs of all 10 ARMD projects in the context of external constraints such as the involvement of center directors in staffing decisions.

The *Decadal Survey of Civil Aeronautics* (NRC, 2006) reported that ARMD had plans to allocate only 7 percent of its budget for research by outside organizations. The *Decadal Survey* recommended that NASA ensure the substantive involvement of universities and industry in civil aeronautics research, in part by establishing a more balanced allocation of funding between in-house and external organizations. ARMD has subsequently increased the percentage of research funds assigned to contracts and grants to academia and industry by way of NASA Research Announcements (NRAs) as opposed to in-house research by NASA civil servants. This percentage varies among the projects from 10 percent to more than 40 percent. As the percentage of NRA funds increases, NASA is able to leverage the expertise and capabilities of outside organizations, and NASA personnel may have the freedom to focus more on integration and application of foundational technologies to higher-level systemic issues. However, it is not clear that this will happen. As external research increases, the civil servant workforce will need to dedicate more time to monitor the performance of the NRA contractor versus doing significant research on their own. At least one PI noted that he had to restructure his project organization to accommodate the demand for experienced contract monitors. In essence, the valuable infusion of new technology and innovative research approaches provided by the use of NRAs should be balanced by the retention of sufficient numbers of experienced NASA civil servants to provide for an optimum mix of research personnel resources, in accordance with a strategic workforce plan for NASA aeronautics research.

Most PIs felt that, at least in the short term (2 to 5 years) they had adequate personnel resources available to perform the required research. However, they universally noted that, as the research shifts from civil servants to university- or industry-based NRA contracts, less of the work is done by long-term subject-matter experts and more of the work is being done by bright young graduate students. When a second or third round of NRA contracts is being let, there is no assurance that the competitively selected NRA contractor will use the same personnel as on earlier related NRAs. However, PIs commented that in many skill areas NASA's permanent staff "is one deep," and that the loss of a single individual due to normal job attrition or retirement results in a loss of continuity. There are many potential approaches for dealing with this issue. For example, assuming that the size of the civil servant NASA workforce is fixed, NASA could increase staffing in some areas, at the expense of losing expertise entirely in some other areas. Or, NASA could in some situations resolve to live with the current situation, using NRAs as necessary to fill small gaps in expertise as they arise, at least on a temporary basis. (In some cases, NRA researchers have accepted subsequent employment offers as NASA civil servants in their area of specialty when staffing plans permit, with the NRAs thus serving as a valuable and effective workforce recruiting tool.) Regardless, ARMD should develop a strategy for dealing with this critical issue.

Due to constraints on hiring new NASA civil servants, the NASA workforce strategy of mentoring younger engineers and scientists is not being universally implemented due to the lack of young NASA personnel available to be mentored. It is particularly difficult to recruit new staff at Ames Research Center because the extremely high cost of housing and other living expenses in the area deter young engineers and scientists from moving into the area. In 2006, only seven new civil servants were hired at Ames Research Center—not all of them necessarily researchers. Yet hiring new staff is essential, especially in areas such as Integrated Vehicle Health Management (IVHM), where NASA wants to develop databases and data mining techniques. These skill sets are highly desirable by companies such as Google, which is located very close to Ames Research Center. Because of constraints on hiring and salaries and because of the local competition for key talent, recruiting and developing leaders at NASA Ames in this new area of research might prove challenging.

Another result of constraints imposed by agency personnel rules and practices is the policy of restricting the transfer of staff positions from one NASA center to another.

NASA's aeronautics research is performed almost entirely at four research centers (Ames Research Center, Dryden Flight Research Center, Glenn Research Center, and Langley Research Center). During 2006, the total number of civil servant employees at these four research centers dropped by about 6 percent, while the number of civil servants at the rest of NASA dropped by just 1 percent. Furthermore, from 2004 to 2006, for every five employees who left these four centers, only one new employee was hired, and the total civil servant workforce at these centers fell by 16 percent. Meanwhile, in the rest of NASA, for every five civil servant employees who left, three new employees were hired, and the number of employees declined by just 2 percent. Expressed another way, from 2004 to 2006, NASA's four research centers, with less than one-third of NASA's total civil service workforce, absorbed almost 80 percent of NASA's reduction in civil service employees (BPTW, 2007). This is consistent with NASA's own longer-range plans. According to NASA's *Workforce Strategy*, from fiscal year (FY) 2005 to FY 2011, the civil servant workforce at NASA's four research centers will be reduced by 19 percent, while the workforce at the rest of NASA is reduced by 5 percent (NASA, 2006b). The *Workforce Strategy* notes that "changes in workyear requirements through fiscal year 2010 are primarily driven by continuing redeployment of the workforce, especially to effect the restructuring of the aeronautics program and the development and testing of the Space Shuttle follow-on systems."

The committee did not have employment data broken down by job category (administrative, technical, etc.), nor did it have data on the size of NASA's total aeronautics research workforce, which includes contractor and academic staff funded by NASA to conduct aeronautics research. Nevertheless, the civil servant staffing trends noted above are both a challenge and an opportunity. On the one hand, lower civil servant staffing may make it more difficult for NASA to conduct meaningful in-house research and interact in a meaningful way with outside researchers. Low hiring rates, even in the face of normal attrition, also make it more difficult to hire significant numbers of younger researchers. On the other hand, lower civil servant staffing (and reduced staffing costs) may free up resources to involve more outside researchers in contributing to the accomplishment of ARMD research goals.

Conclusions

A recent National Research Council report, *Building a Better NASA Workforce* (NRC, 2007), shows that much of the aerospace workforce on which NASA's space program has historically relied and will continue to rely exists outside the agency in industry or universities. In addition, even though the average age of NASA's workforce has steadily increased over the past 15 years or so (to the point where perhaps 25 percent of NASA's workforce could retire in the next 5 years), this problem is not unique to NASA. In fact, the federal government as a whole is much more susceptible than NASA is to a mass exodus of employees as a large percentage of its aging workforce becomes eligible to retire. *Building a Better NASA Workforce* also concludes that "the general focus on the age of the NASA workforce and a looming 'retirement crisis' tends to obscure more complex and subtle demographic issues. Although a massive and simultaneous wave of retirements among eligible employees would be a devastating blow to the agency, it is likely that NASA will continue to retain employees beyond retirement age and to engage the retiree community as consultants and mentors, as it has done in the past" and "the most relevant issue facing NASA's workforce is not its age, but rather the number and distribution of skilled employees within the agency and the ability of the agency to ensure that it has, and will continue to have, an adequate supply of trained employees" (NRC, 2007, p. 10).

Specifically with regard to aeronautics, this committee found that there exists, among NASA, the academic community, and the civilian aerospace industry, enough skilled research personnel to adequately support the current aeronautics research programs at NASA and nationally, at least for the next decade or so. NASA may experience localized problems at some centers. For example, Ames Research Center may have difficulty hiring civil servants to do some work in-house. However, the requisite intellectual capacity exists at other centers and/or in organizations outside NASA. Thus, NASA should be able to achieve its research goals, for example, by using NRAs or other procurement mechanisms; through the use of higher, locally competitive salaries in selected disciplines at some centers; and/or by creating a virtual workforce that integrates staff from multiple centers with the skills necessary to address a particular research task.

The content of the NASA aeronautics program, which has a large portfolio of tool development but little or no opportunities for flight tests, may in some cases hamper the ability of the aeronautics program to recruit new staff as compared with the space exploration program. In addition, there will likely be increased requirements for specialized or new skill sets, such as IVHM (as discussed above) and system engineering expertise (to facilitate management of a growing portfolio of research performed by external organizations).

Workforce problems and inefficiencies can also arise from fluctuations in national aerospace engineering employment and from uneven funding in particular areas of endeavor. If expertise is developed when the aerospace industry is growing and/or funding for a particular program is available, and if that expertise is lost when the industry contracts, when programs and contracts end, or when budgets shift focus, then research teams disburse into other endeavors, sometimes in other industries. As an extreme example, NASA has recently discovered that the ability to fabricate ablative material for the heat shields of Apollo command modules returning from lunar orbit has been lost. In fact, heat shield test articles fabricated according to available records from the Apollo program failed almost immediately during testing, presumably because the records do not document some vital aspect of the fabrication process.

NASA's ability to attain its aeronautics research goals will be greatly facilitated by the expertise of the staff who are leading and conducting ARMD's aeronautics research projects—and who are dedicated to the success of these projects.

Recommendation. To ensure that the NASA aeronautics program has and will continue to have an adequate supply of trained employees, the Aeronautics Research Mission Directorate should develop a vision describing the role of its research staff as well as a comprehensive, centralized strategic plan for workforce integration and implementation specific to ARMD. The plan should be based on an ARMD-wide survey of staffing *requirements* by skill level, coupled with an *availability* analysis of NASA civil servants available to support the NASA aeronautics program. The plan should identify specific gaps and the time frame in which they should be addressed. It should also define the role of NASA civil servant researchers vis-à-vis external researchers in terms of the following:

- Defining, achieving, and maintaining an appropriate balance between in-house research and external research (by academia and industry) in each project and task, recognizing that the appropriate balance will not be the same in all areas.
- Maintaining core competencies in areas consistent with (1) the highest-priority R&T challenges from the *Decadal Survey of Civil Aeronautics* and (2) NASA's role in the *National Aeronautics*

Research and Development Policy and the *National Plan for Aeronautics Research and Development and Related Infrastructure*.
- Supporting the continuing education, training, and retention of necessary expertise in the NASA civil servant workforce and, as appropriate, determining how to encourage and support the education of the future aeronautics workforce in general.
- Developing, integrating, and applying foundational technology to meet NASA's internal requirements for aeronautics research.
- Defining and addressing issues related to research involving multidisciplinary capabilities and system design (i.e., research at Levels 3 and 4, respectively, as defined by ARMD).
- Ensuring that research projects continue to make progress when NASA works with outside organizations to obtain some of the requisite expertise (when that expertise is not resident in NASA's civil servant workforce).

NASA should use the National Research Council report *Building a Better NASA Workforce* (NRC, 2007) as a starting point in developing a comprehensive ARMD workforce plan.

AERONAUTICS FACILITY ISSUES

Aeronautics Test Program Facilities

ARMD's Aeronautics Test Program (ATP) is responsible for providing corporate management of NASA's major aeronautical facilities. This includes increasing the probability of having the right facilities in place at the right time over the long term, operating facilities in the most effective and efficient manner possible, and ensuring intelligent investment in and divestment of NASA facilities. The ATP supports operations (facility sustainment and rate stabilization), maintenance, investments in test technology, and university-related research. ATP facilities are located at Ames Research Center, Glenn Research Center, Langley Research Center, and Dryden Flight Research Center. To be included in the ATP, facilities must meet a size criterion and provide unique capabilities. ATP ground test facilities include the following:

- Ames Unitary Wind Tunnel
- Glenn Icing Research Tunnel
- Glenn 9- × 15-Foot Subsonic Tunnel
- Langley National Transonic Facility
- Langley Transonic Dynamics Tunnel
- Langley Hypersonic Complex
- Langley 8-Foot High Temperature Tunnel
- Langley 14- × 22-Foot Subsonic Tunnel
- Langley 20-Foot Vertical Spin Tunnel
- Glenn Propulsion Systems Laboratory 3 and 4
- Glenn 10- × 10-Foot Supersonic Tunnel

ATP flight research facilities include the following:

- Western Aeronautical Test Range
- Support Aircraft
- Test Bed Aircraft
- Simulation and Flight Loads Laboratory

The ATP continually assesses the need for its facilities and, as appropriate, reduces the number of active facilities, in part to eliminate duplicate facilities. The ATP currently has two mothballed facilities: the Glenn Hypersonic Test Facility and the Ames 12-Foot Subsonic Pressure Tunnel. The Langley Low Turbulence Pressure Tunnel is being considered for deactivation. Six facilities are scheduled for demolition: the Ames 14-Foot Transonic Facility, the Langley 7- × 10-Foot High Speed Tunnel, the Langley 16-Foot Transonic Tunnel, Langley's two 8-Foot Transonic Tunnels, and the Langley 30- × 60-Foot Full Scale Tunnel. Although mothballing and demolishing facilities may seem drastic, NASA has carefully considered how such actions would affect current and future aeronautics research, and the process seems to be working well.

Over the past several years, the ATP annual budget has been approximately $175 million. About $87 million has been recovered from users who reimburse NASA for direct costs associated with their tests. About 68 percent of this amount has come from NASA users (of which 60 percent has come from the Fundamental Aeronautics Program). The rest of ATP's annual budget (about $88 million) has come from NASA's congressionally appropriated funds through the ATP. These funds have been allocated as follows: $31 million for flight operations and test infrastructure and $57 million for aeronautics ground test facilities ($39 million for operations and $18 million for maintenance and test technology, including university-related research). In addition, Center Management and Operations funds have provided approximately $14 million per year of reactive and preventive maintenance for ATP facilities. In other words, the total budget for the maintenance and improvement of ATP ground facilities has been approximately $32 million per year ($18 million from the ATP budget and $14 million from the centers' budgets).

The ATP staff consists of a small program office, lead personnel at each NASA center with ATP facilities, and remaining staff to accomplish the testing, maintenance, and investments funded by the program.

The ATP collaborates closely with the Department of Defense (DoD) though the recently signed National Partnership for Aeronautical Testing (NPAT). The NPAT is intended to increase communications between DoD, NASA, and industry in the area of aeronautical test facilities. It encourages NASA and DoD to address the nation's aeronautical test needs with less duplication by relying on each other for certain capabilities.

Shared Capability Assets Program Facilities

The Shared Capabilities Assets Program (SCAP) within the NASA Headquarters Infrastructure Office identifies, prioritizes, and supports key facilities that NASA deems essential to the future needs of NASA and/or the nation. In some cases NASA provides funding to maintain critical capabilities that are underutilized at present so they will still be available when they are needed. The SCAP has in its portfolio the kinds of facilities listed below. As shown, some are managed by SCAP, and some are managed by NASA's mission directorates.[3]

- Aeronautics Test Program (managed by ARMD)
- Arc jet test facilities (managed by SCAP)
- Flight simulation facilities (managed by SCAP)
- High-End Computing Capability (managed by the Science Mission Directorate)

[3] In general, facilities managed by individual directorates primarily support those directorates, whereas facilities managed by SCAP are not as closely linked to any one directorate.

- Rocket Propulsion Test Program (managed by the Space Operations Mission Directorate)
- Thermal-vacuum-acoustic test facilities (managed by SCAP)

All but the last two support aeronautics research and development.

The ATP facilities budget is described in detail above. The approximate SCAP budget for the other capabilities that support aeronautics is as follows:

- *Arc jet:* $14 million per year ($9 million for operations support and maintenance and $5 million from customer reimbursement for the marginal costs of testing).
- *Flight simulation:* $15 million per year ($11 million is for operations support and maintenance and $4 million from customer reimbursements).
- *High-end computing:* $39 million per year ($23 million for operations support, $6 million for maintenance, and $10 million for capital improvements).

All customers currently pay the same rates for using the three SCAP-managed facilities, but this policy is under review.

Other Laboratory Facilities

ARMD's aeronautics research also uses other small research laboratories located at various centers. These laboratories are sustained and maintained by the individual centers (using center Management and Operations funds), or they are directly funded by the ARMD program requiring their use. Examples include combustor research facilities, fuel research facilities, and crash facilities. In addition, the projects within the Fundamental Aeronautics Program invest in research to provide new and/or improved test techniques and instrumentation, as necessary.

Requirements for NASA Research Facilities

NASA Aeronautics Program

NASA's aeronautics research uses a wide variety of facilities. For example, ARMD's Airspace Systems Project uses SCAP-funded facilities such as the Crew Vehicle Systems Research Facility, the Cockpit Motion Facility, and the Future Flight Central (Tower Cab). Both the Airspace and the Airportal Projects use the North Texas Research Station, which is supported by NASA and the Federal Aviation Administration (FAA). (This is a small facility that supports field testing of automated air traffic management tools.)

The Integrated Resilient Aircraft Control Project uses flight simulation facilities, full-scale flight vehicles, and subscale flight vehicles (such as the scale model transport testbed).

The Subsonic Rotary Wing Project utilizes the Langley 14- × 22-Foot Tunnel and the Langley Transonic Dynamics tunnel.

The Supersonics Project utilizes the Glenn 10- × 10-Foot Tunnel, the Glenn 1- × 1-Foot Tunnel, the Langley Unitary Plan Wind Tunnel, and the Ames Unitary Plan Wind Tunnel.

The Hypersonics Project uses the Langley hypersonic facilities.

NASA Space Programs

The NASA Science Mission Directorate is developing the Mars Science Laboratory. This effort includes two options for addressing issues associated with vehicle entry, descent, and landing. One option, which would use aerodynamic drag deceleration, requires arc-jet facilities to test candidate heat shields. The second option, which would use supersonic parachutes, requires access to NASA's large supersonic wind tunnels.

NASA's Exploration Systems Mission Directorate is working on the Project Constellation, which includes robotic and human exploration of the Moon and Mars. Constellation's reentry vehicle, Orion, will have more mass than previous reentry vehicles and will therefore require the development of advanced entry, descent, and landing systems. Project Constellation will be responsible for funding necessary improvements to existing arc-jet facilities in order to support this development. Supersonic parachutes are also being considered for Orion and will need supersonic wind tunnel support. Project Constellation will also use ATP wind tunnels to support development of the Ares launch vehicle, particularly with regard to launch vehicle stability and control, buffet response and loads, structures, stage separation, heating, and the performance of the launch abort vehicle.

The Space Operations Mission Directorate utilizes NASA's aeronautics facilities to support the space shuttle operations. Accident Investigation Boards use facilities to better understand failure mechanisms, and program staff use facilities to test fixes and modifications during the process of returning to flight.

DoD Programs

The National Defense Authorization Act for Fiscal Year 2005 requested that DoD identify and analyze NASA aeronautics facilities that are critical to defense missions. In response, DoD identified 11 NASA aeronautics facilities whose continued availability is necessary to avoid an unacceptable risk to research, development, modernization, and sustainment of the weapon systems supporting the defense mission (DoD, 2007). These facilities are as follows:

- Glenn Research Center 6- × 9-Foot Icing Research Tunnel
- Langley Research Center 20-Foot Vertical Spin Tunnel
- Ames Research Center Unitary 11-Foot Transonic Tunnel
- Langley Research Center National Transonic Facility
- Langley Research Center Transonic Dynamics Tunnel
- Langley Research Center 8-Foot High Temperature Tunnel
- Ames Research Center Vertical Motion Simulator
- Glenn Research Center Mechanical Drives Facility
- Glenn Research Center Turbine and Structural Seals Facilities
- Langley Research Center Impact Dynamics Research Facility
- Wallops Flight Facility Open Air Range

Some of these facilities are critical to DoD because they have unique capabilities. Others are critical because the workload is so high that other, comparable facilities cannot meet the combined workload of DoD, NASA, and the U.S. aerospace industry. And some of the above facilities are critical because other comparable facilities may experience a lengthy period of unplanned or planned downtime (due to failure, natural catastrophe, or modification), and a second facility is necessary because the consequences of deferring relevant research and testing are unacceptable.

The DoD study identified the following specific requirements for ATP facilities:

- The Icing Research Tunnel at Glenn is a core facility that is critical to manned and unmanned aircraft and pilot survivability in severe icing conditions for all flight profiles. All of the military services have a potential need for this facility to provide long-range support during system development, conduct credible large-scale examinations of icing phenomenology and ice protection systems, and generate tools needed for ice protection systems in future aircraft designs.
- The Vertical Spin Tunnel at Langley is the largest spin tunnel in the world, it is the only free-spin tunnel in the United States, and it is essential to determining certain aerodynamic characteristics of the highly maneuverable aircraft that are developed and modified by the DoD.
- The transonic regime is both complex and critical to weapon systems development. Due to the high workload in this test regime, the DoD needs to maintain access to more than one large-scale transonic wind tunnel. The combination of test article size, Mach number, and altitude range of the Unitary 11-Foot Transonic Wind Tunnel at Ames makes it critical to DoD transonic testing requirements for aerodynamic simulation, especially for long-range aircraft systems.
- The National Transonic Facility at Langley is vital to determining scaling effects for transport aircraft, bombers, and other long-range vehicles. To assure that modeling and simulation tools can accurately extrapolate Reynolds number effects far enough to accurately predict flight characteristics, modeling and simulation tools must be calibrated using physical testing under high Reynolds numbers. This testing is the primary purpose of the National Transonic Facility.
- The Transonic Dynamics Tunnel is essential for DoD testing for both flutter clearance and structural aeroelasticity. This facility is necessary to support future heavy-lift rotorcraft designs.
- The High Temperature Tunnel at Langley provides DoD with hypersonic aerothermal and air-breathing propulsion environments for Mach 4 to 7 systems while the DoD's Aerodynamic and Propulsion Test Unit facility is being upgraded. In addition, due to the high workload in this test regime, the DoD needs to maintain access to more than one hypersonic aerothermal and aerodynamic test environment.

FAA Programs

The Federal Aviation Administration has used several NASA facilities to support past research, and this need will continue, especially with regard to NextGen research.

Future Flight Central, a SCAP-managed facility at Ames, is a national air traffic control/air traffic management test facility dedicated to solving present and future capacity problems at U.S. airports. This facility has established a precedent for enabling stakeholders to achieve consensus through a common vision of the future.

The Crew-Vehicle Systems Research Facility is another SCAP-managed facility at Ames. This unique facility supports research by NASA, the FAA, and industry. This facility provides an environment where researchers can study how and why aviation errors occur, and it stands out in the area of human factors research. In this facility, researchers use highly sophisticated flight and air traffic control simulators to study the effects of automation and advanced instrumentation on human performance.

The Air Traffic Operations Laboratory at Langley is a center-managed facility that concentrates more on the air traffic control issues than the Crew-Vehicle Systems Research Facility. This facility includes controller displays that depict data generated by either traffic generator software or by integrating flight simulators from Langley or other facilities around the country.

The FAA has also made use of other facilities, such as the Cockpit Motion Facility and the Vertical Motion Simulator (which are managed by SCAP) and the Landing Loads and Crash facility (which is managed as a center laboratory).

None of the above facilities needs to be upgraded to meet the FAA's current research requirements. The ability of these facilities to support future requirements is uncertain, given that the FAA research requirements related to NextGen are still evolving.

Industry Programs

The American Institute of Aeronautics and Astronautics (AIAA) U.S. Industry Aeronautics Test Facilities Working Group provides a forum for development of strategic recommendations on aeronautical wind tunnels required to support current and future aeronautics research and development. Recently the working group developed a position statement addressing infrastructure recommendations for implementation of a national aeronautics research and development policy (AIAA, 2007). This paper addresses total aeronautical testing needs for industry, and it includes some requirements for NASA facilities. The working group evaluated historical wind tunnel test usage trends and estimated industry needs over the next 5 years. Due to the highly cyclical nature of wind tunnel testing (large peaks and valleys in test requirements), the working group used an average industry usage over the past 5 years to produce a baseline. The working group concluded that "wind tunnel testing needs will continue at rates slightly above the five-year annualized baseline. Historical peaks and valleys, resulting from individual program plans, will continue to impact actual requirements and schedules. The five-year forecast indicates that subsonic and transonic testing will remain constant or possibly increase slightly; there will likely be increased emphasis on supersonic and hypersonic testing" (AIAA, 2007).

The working group also evaluated longer-term trends to develop a vision of the future for wind tunnel testing and concluded that "wind tunnel testing is anticipated to continue to provide a large percentage of development and validation data needed in the pursuit of new technologies and systems of aeronautic vehicles. Aeronautic development requires productive and capable tools, so wind tunnels and computational methods will continue to provide the bulk of required data in the future. Aeronautic vehicle development will continue to push the boundaries of our knowledge, increasing the need for tools that can accurately (and efficiently) predict aeronautic effects." The working group identified five strategic investment areas in wind tunnel test capabilities that are required to enable continued progress in aeronautics:

1. Development of a knowledgeable test workforce is critical for the national infrastructure. It is anticipated that the industry-wide trend of losing expertise will particularly impact the test community. Investment, in the form of stable research and development programs (federally funded) which use the test infrastructure, will significantly enhance the ability of the operators to seek and retain new staff prior to the retirement of the current aging workforce. Stable facility funding profiles (e.g. overheads, maintenance, and test technology development), will also aid in developing and retaining knowledgeable staffs.

2. Improved test technology is crucial to enabling future system development. Huge advances in the ability to mine flow field and other data from wind tunnel tests are possible (through the advance in computational capabilities and integration, instrumentation and non-intrusive flow field measurement techniques) with strategic and coordinated investment. Stable funding of test technique/technology research would also support the development of the test workforce, as described in item 1 above.

3. Maintenance and improvement of key test assets is a vital component of enabling future test capabilities. Key facilities include those that provide unique, or nearly unique, capabilities and may or may not have

full utilization. Atrophy and decline in a robust Federal aerodynamic test infrastructure in the last twenty years has been significant; only recent emphasis by the DOD and NASA has slowed the decline in our aging infrastructure.

4. Divestment of redundant and non-essential test infrastructure is required to focus limited resources on critical capabilities and new infrastructure requirements.

5. New high-speed test infrastructure is required to meet anticipated requirements for future systems. Increased need for simulation of hypersonic realm will likely require additional test capabilities not currently available. (AIAA, 2007, p. 9)

NASA Research Facility Gaps

Principal investigators for ARMD's 10 research projects did not identify any serious capability gaps in the NASA research facilities needed to support their research, except for one concern with NASA's quiet supersonic tunnel capability. A quiet supersonic wind tunnel with maximum operating conditions of about Mach 1.5 to 2.5 is needed to validate designs as flows transition between turbulent and laminar boundary layers.

NASA's Exploration Systems, Space Operations, and Science Mission Directorates identified no unmet facility needs other than improved arc-jet facilities.

The DoD and FAA reported that the current NASA capabilities are doing an excellent job of meeting relevant testing requirements.

Industry forecasts continued reliance on NASA aeronautical facilities to meet future research requirements, and it predicts that new high-speed infrastructure will be required to meet future requirements.

However, the question that all facility users are asking is whether NASA is investing in and maintaining its facilities at a level that will ensure that the current status will be continued into the future. The analysis below indicates that it is not.

In FY 2006 NASA performed its fifth facilities condition assessment and deferred maintenance study (NASA, 2006a). This assessment determined that the NASA-wide facility condition index, which is rated on a scale from 5 (excellent) to 1 (bad), had decreased slightly from 3.7 in FY 2005 to 3.6 in FY 2006. This means that NASA's facilities remain in fair condition, meaning they are "occasionally unable to function as intended." Center staff has been able to maintain this condition with consistent maintenance and increased use of technology for tracking maintenance and repairs. However, NASA's stated goal is to increase the facility condition index to 4.3. As described below, this is unlikely to happen with the current level of maintenance funding. In particular, ATP facilities are likely to degrade over time unless the inventory of ATP facilities is substantially reduced or funding for facilities is increased.

In 1989 the Institute for Defense Analyses (IDA) assessed the adequacy of DoD expenditures to replenish the capital stock at the 22 principal test and evaluation (T&E) sites that constitute the DoD's Major Range and Test Facility Base (IDA, 1989). The capital stock of interest included buildings, towers, test stands, range instrumentation, data processing equipment, and other structures and major equipment. The capital stock at the sites examined by this study would be very similar to NASA's current ATP facilities. The study compared the rate of capital renewal at the 22 DoD T&E sites with the rate at major industries, nine principal aerospace firms, and nonmilitary federal departments and agencies. The capital renewal period was obtained by dividing the value of capital stock by the level of annual investment in existing and new facilities. The renewal period indicates how long it would take to completely replace the capital stock at current levels of investment. The longer the renewal period, the older the stock of buildings and equipment would become in the future. The IDA report found that the DoD T&E sites had a renewal period of 64 years, which was much longer than the capital renewal periods for non-DoD federal departments and agencies (22 years), aerospace firms (18 years), and major industry as a whole (14 years).

The IDA (1989) report caused the DoD to make significant changes to reduce the renewal period of its T&E sites. In the mid-1990s the DoD established the Central Test and Evaluation Investment Program, funded at $140 million per year to reduce the renewal period. Later, the Fiscal Year 2003 National Defense Authorization Act directed the DoD and the military services to fully fund the institutional and overhead costs (including maintenance) of its T&E sites and only charge users the direct costs associated with their tests. Also in 2003, the U.S. Air Force increased the maintenance budget at the Arnold Engineering Development Center by $15 million per year to help reduce the backlog of maintenance and repair projects at the center. These efforts have substantially reduced the renewal period for major DoD T&E facilities.

The NASA Real Property Database indicates that the capital stock value of ATP ground test facilities as of 2005 is $1.9 billion, based on the original construction cost, adjusted for inflation. This may understate the true cost of replacing these assets, in part because substantial improvements have been made over the years using civil servant labor, the cost of which is not reflected in the NASA Real Property Database. Therefore, $1.9 billion is a conservative estimate of value of ATP facilities. Dividing that number by the annual maintenance and investment budget for ATP facilities ($32 million), produces a renewal period of 59 years. This indicates that NASA's situation today is similar to DoD's situation in 1989, and NASA and Congress may need to take similar corrective action to prevent the status of "occasionally unable to function as intended" from declining to "often being unable to function." Such an outcome would impede important research by NASA, DoD, FAA, and industry.

Conclusions

NASA has a unique set of aeronautics research facilities that provide key support to NASA, other federal agencies, and industry. With very few exceptions, these facilities meet the relevant needs of existing aeronautics research. NASA also has a dedicated effort to sustain large, key facilities and to shut down low-priority facilities. However, some small facilities (particularly in the supersonic regime) are just as important and may warrant more support than they currently receive. In addition, at the current investment rate, widespread facility degradation will inevitably impact the ability of ARMD projects and other important national aeronautics research and development to achieve their goals.

Recommendation. Absent a substantial increase in facility maintenance and investment funds, NASA should reduce the impact of facility shortcomings by continuing to assess facilities and mothball or decommission facilities of lesser importance so that the most important facilities can be properly sustained.

REFERENCES

AIAA (American Institute of Aeronautics and Astronautics). 2007. Infrastructure Recommendations for Implementation of Executive Order 13419, National Aeronautics Research and Development: An AIAA Position Statement. Reston, Va.: American Institute of Aeronautics and Astronautics. Available online at <http://pdf.aiaa.org/downloads/publicpolicypositionpapers//windtunnelinfrastucpaperbodapproved011108.pdf>.

BLS (Bureau of Labor Statistics). 2006. Bureau of Labor Statistics, U.S. Department of Labor, Occupational Outlook Handbook, 2006-07 Edition, Engineers. Washington, D.C.

BLS. 2007. Median weekly earnings of full-time wage and salary workers by detailed occupation and sex, 2006. Available online at <www.bls.gov/cps/cpsaat39.pdf>.

BPTW (Best Places to Work). 2007. Partnership for Public Service and American University's Institute for the Study of Public Policy Implementation (ISPPI). Available online at <http://bestplacestowork.org>.

DoD (Department of Defense). 2007. NASA Aeronautics Facilities Critical to DoD Report to Congress. Under Secretary of Defense for Acquisition, Technology and Logistics. Washington, D.C.: Department of Defense. Available online at <www.acq.osd.mil/trmc/publications/NASA%20Report%20to%20Congress.pdf>.

IDA (Institute for Defense Analyses). 1989. Maintaining the Capital Stock at Department of Defense Test and Evaluation Sites. Alexandria, Va.: Institute for Defense Analyses.

NAPA (National Academy of Public Administration). 2005. NASA: Human Capital Flexibilities for the 21st Century Workforce. Washington, D.C.: National Academy of Public Administration.

NASA (National Aeronautics and Space Administration). 2006a. Deferred Maintenance Assessment Report, FY06 NASA-Wide Standardized Deferred Maintenance Parametric Estimate—Full Assessment, Final. Washington, D.C.: National Aeronautics and Space Administration.

NASA. 2006b. National Aeronautics and Space Administration Workforce Strategy. Washington, D.C.: NASA. Available online at <http://nasapeople.nasa.gov/HCM/WorkforceStrategy.pdf>.

NRC (National Research Council). 2006. Decadal Survey of Civil Aeronautics: Foundation for the Future. Washington, D.C.: The National Academies Press. Available online at <http://www.nap.edu/catalog.php?record_id=11664>.

NRC. 2007. Building a Better NASA Workforce: Meeting the Workforce Needs for the National Vision for Space Exploration. Washington, D.C.: The National Academies Press. Available online at <www.nap.edu/catalog.php?record_id=11916>.

White House. 2004. President Bush Announces New Vision for Space Exploration Program. Washington, D.C.: White House. Available online at <http://www.whitehouse.gov/news/releases/2004/01/20040114-3.html>.

4

Bridging the Gaps

This chapter identifies key gaps that must be eliminated for NASA's aeronautics program to meet key goals in terms of the research and technology (R&T) challenges from the *Decadal Survey of Civil Aeronautics* (NRC, 2006) as well as internal NASA requirements for aeronautics research and the requirements that NASA is expected to satisfy in support of aeronautics research by other federal agencies.

One of the federal government's goals for its aeronautics research and development (R&D) is to "cultivate a research and development environment that enables a globally competitive U.S. aeronautics enterprise, and encourages industry investment and academic participation" (NSTC, 2006, p. 4). One of NASA's objectives is to preserve "the role of the United States as a leader in aeronautical and space science and technology and in the application thereof to the conduct of peaceful activities within and outside the atmosphere," in part through "research into, and the solution of, problems of flight within and outside the Earth's atmosphere, the development, construction, testing, and operation for research purposes of aeronautical and space vehicles" (National Aeronautics and Space Act of 1958, Public Law No. 85-568, as amended). Consistent with the above, the *Decadal Survey of Civil Aeronautics* recommends that the federal government take action to ensure U.S. leadership in civil aeronautics.

The current approach used by NASA's Aeronautics Research Mission Directorate (ARMD) for achieving agency goals related to aeronautics is embodied in the following principles (Porter, 2007, p. 9):

- We will dedicate ourselves to the mastery and intellectual stewardship of the core competencies of aeronautics for the nation in all flight regimes.
- We will focus our research in areas that are appropriate to NASA's unique capabilities.
- We will directly address the fundamental research needs of the Next Generation Air Transportation System (NGATS/NextGen)[1] in partnership with the member agencies of the Joint Planning and Development Office (JPDO).

[1]Next Generation Air Transportation System, formerly known by the abbreviation NGATS, is now more commonly referred to as NextGen.

As noted in the statement of task for this study, NASA uses the term *fundamental research* to include continued long-term, scientific study in areas such as physics, chemistry, materials, experimental techniques, and computational techniques that leads to a furthering of the understanding of the underlying principles that form the foundation of the core aeronautics disciplines, as well as that research that integrates the knowledge gained in these core areas to significantly enhance NASA's capabilities, tools, and technologies at the disciplinary (e.g., aerodynamics, combustion, trajectory prediction uncertainty) and multidisciplinary (e.g., airframe design, engine design, airspace modeling and simulation) levels.

To effectively support U.S. efforts to maintain a position of aeronautics leadership and competitiveness, NASA research at all levels must be carefully structured in terms of work content, performance expectations, and technology transfer to external and internal users of NASA aeronautics research, which include U.S. industry, the Federal Aviation Administration (FAA), Department of Defense (DoD), academia, and the NASA space program. Foundational research that addresses underlying principles will tend to be earlier in the technology maturity continuum, and it will be more scientific in nature, which makes it more difficult to schedule meaningful milestones and to point this research in a direction that is sure to enhance U.S. competitiveness and meet user needs. On the other hand, more advanced research with a clear path to application will likely have a heightened sense of urgency and purpose. The ideal plan would have a combination of both foundational and applied research in appropriate areas, with management systems appropriate for each. For example, integration of multidisciplinary research related to the health of vehicle systems and vehicles as a whole (see R&T challenge D5) would be more effective if managed through an overarching and formal organizational entity that relies on system engineering disciplines.

Maintaining a position of leadership requires staying ahead of the pack, by being the first to bridge each new gap into the future. This is a challenging task; were it not so, others would have overtaken the leader to set a faster pace. The statement of task for this study directs the committee to consider whether there are gaps, and "if gaps are found, what steps should be taken by the federal government to eliminate them." Looking to the future of NASA's aeronautics research, the following gaps will require special attention to ensure that the nation's civil aeronautics research program, executed through NASA, achieves its goals.

GAP BETWEEN RESEARCH RESULTS AND APPLICATION

The *Decadal Survey of Civil Aeronautics* evaluates R&T challenges based on their potential to achieve six strategic objectives:

- Increase the capacity of the air transportation system.
- Improve the safety and reliability of the air transportation system.
- Increase the efficiency and performance of aircraft, facilities, and so on to maximize utilization of the air transportation system.
- Reduce energy consumption and the negative environmental impact of air transportation.
- Take advantage of synergies when specific aeronautical research helps to achieve the first four objectives while also helping to achieve the goals of the DoD and the Department of Homeland Security.
- Support the space program.

Maintaining core competencies, especially in areas that resonate with NASA's unique capabilities, is an appropriate part of NASA's strategy for maintaining a U.S. position of leadership in aeronautics. However, NASA does not manufacture, own, or operate the aircraft and ground systems that comprise the air transportation system. Thus, for NASA research to provide value to its stakeholders and/or achieve the performance-enhancing goals of the *Decadal Survey of Civil Aeronautics,* its research results must ultimately be transferred to industry, the FAA, and the other organizations that do manufacture, own, and operate key elements of the air transportation system. For example, the projects that are working on R&T challenges related to the air transportation system should articulate, at some level, how research results are tied to capability improvements as well as a transition path that will help guide researchers in planning and conducting research in such a way that it is likely to be of value to external users. Furthermore, projects that are developing modeling and analysis tools should have a plan for validating the tools in particular applications of interest, such as investigating the performance of one or more notional aircraft.

Technology transfer issues are especially critical given that NASA's current approach to aeronautics research will not mature technology as far as in the past, and this will require users, such as the FAA and industry, to adjust their expectations and processes accordingly.

NASA's legislative charter (the National Aeronautics and Space Act of 1958, as amended) establishes "the preservation of the role of the United States as a leader in aeronautical and space science and technology" as one of NASA's objectives. In addition, the *National Aeronautics Research and Development Policy* (NSTC, 2006) confirms that "a continued strong U.S. government role in aeronautics R&D is needed to . . . create an environment in which U.S. industry remains innovative and competitive." As noted above, ARMD has dedicated itself to the mastery and intellectual stewardship of core competencies in all flight regimes. Yet this is only a means to an end, and the larger purpose is NASA's legislative mandate to ensure continued U.S. leadership in aeronautics. NASA's ability to fulfill this mandate would be jeopardized if NASA were to view itself as the ultimate beneficiary of its own aeronautics research. This outcome will be avoided if NASA's efforts to maintain core competencies are structured to achieve the larger goals of (1) maintaining U.S. leadership and industrial competitiveness, (2) accomplishing the strategic objectives of the *Decadal Survey of Civil Aeronautics,* and (3) fulfilling the principles described in the *National Aeronautics Research and Development Policy* (NSTC, 2006) and the *National Plan for Aeronautics Research and Development and Related Infrastructure* (NSTC, 2007). This difficult task would be facilitated if the principal investigators (PIs) for NASA's aeronautics research projects identify and connect with the organizations that will bridge the gap between NASA research results, user needs, and implementation of the larger goals. In addition, for technology intended to enhance the competitiveness of U.S. industry, U.S. leadership would be enhanced by a technology transfer process that does not necessarily include immediate, public dissemination of results to potential foreign competitors, so that the U.S. industrial base has a head start in absorbing the fruits of this research. This is not necessary for noncompetitive research, such as investigating the impact that aviation plays on the environment, or other research that NASA may conduct as part of international collaborations. In addition, the quality of NASA's research would be enhanced by technology transfer processes designed to (1) accommodate external peer review as much as possible and (2) avoid locking academia out of the NASA aeronautics program (given academia's understandable interest in open publication of results by its researchers).

A closer connection between NASA aeronautics PIs and some potential users of their research would also facilitate the formation of a coordinated set of research goals and milestones that are timed to anticipate user needs. A closer examination of NASA's research into airspace issues serves to illustrate this point. The mostly likely path for research by the NGATS Air Traffic Management (ATM)-Airportal and ATM-Airspace Projects to have a real-world impact is by supporting the JPDO's goals for NextGen. Establishing a close link between the plans for these two projects and the needs of the JPDO was complicated by the fact that a NextGen R&D requirements document was not available when NASA's project plans were being prepared. Even so, initial planning of the NGATS ATM-Airportal and ATM-Airspace Projects represented a good-faith effort to meet the expected NextGen requirements in both content and timing.

Key officials at the U.S. Air Force, the Defense Advanced Research Projects Agency, and the NASA space program seem well connected with and satisfied that the PIs responsible for research related to their areas of concern are aware of their interests and provide necessary and timely support. In addition, the JPDO provides a multiagency forum for improving connections between researchers and users.

Finding. Maintaining technical expertise and core competence in appropriate areas is necessary to maintaining leadership, but it is not sufficient. Close contact with potential users is also necessary, to facilitate research planning and the transfer of research results from NASA to users.

Recommendation. The NASA Aeronautics Research Mission Directorate should bridge the gap between research and application—and thereby increase the likelihood that this research will be of value to the intended users—as follows:

- Foster closer connections between NASA principal investigators and the potential external and internal users of their research, which include U.S. industry, the Federal Aviation Administration, the Department of Defense, academia, and the NASA space program.
- Improve research planning to ensure that the results are likely to be available in time to meet the future needs of the nation.
- Consistently articulate during the course of project planning and execution how research results are tied to capability improvements and how results will be transferred to users.
- For technology intended to enhance the competitiveness of U.S. industry, establish a more direct link between NASA and U.S. industry to provide for technology transfer in a way that does not necessarily include the immediate, public dissemination of results to potential foreign competitors.

As part of the effort to implement this recommendation, NASA should ensure that the NGATS ATM-Airportal and ATM-Airspace Projects meet the R&D needs defined by the NextGen JPDO for NASA.

GAP BETWEEN RESEARCH SCOPE AND RESOURCES

As noted in Chapter 2, NASA is doing a mixed job in responding to the R&T challenges from the *Decadal Survey*. The committee found no significant shortcomings in NASA's efforts to address 4 R&T challenges, and 8 challenges were uniformly evaluated as demonstrating minor shortcomings. For 7 other R&T challenges, NASA is making little or no progress or, even if planned research is successful, it is highly unlikely to make a significant difference in a time frame of interest to users of the research results. For the 32 other R&T challenges, NASA is effectively addressing some areas, but not others, and the overall assessment of these challenges is best described as "mixed." This is not surprising, given that the *Decadal Survey* was not chartered to quantify the cost of achieving the highest-priority R&T challenges, individually or as a whole, nor was it tasked with assessing how those costs might compare to NASA's likely budget for aeronautics research. As a result, NASA simply doesn't have the resources to address all of the 51 highest-priority R&T challenges simultaneously, and attempting to do so would reduce the effectiveness of NASA's aeronautics research. Without additional resources, the value added by the NASA aeronautics program would be enhanced by redefining the scope and priorities of the program, even if all of the 51 highest-priority R&T challenges from the *Decadal Survey of Civil Aeronautics* are not addressed simultaneously. Decisions about which research tasks to pursue in the near term should be guided by the ability of proposed research to meet specified metrics, such as those used in the *Decadal Survey* to identify which challenges were most appropriate for NASA. Those metrics are as follows:

- Supporting infrastructure, which refers to whether NASA already possesses the necessary facilities, resources, and expertise to conduct proposed research.
- Mission alignment, which refers to whether the proposed research falls under NASA's charter, as defined in the National Aeronautics and Space Act of 1958 (as amended).
- Lack of alternative sponsors, which refers to whether other sponsors are able and willing to perform the necessary research. NASA should not be repeating research that is (or should be) done by industry, other federal agencies, or other organizations.
- Appropriate level of risk, which refers to whether the level of risk associated with a research task is appropriate for a NASA research project. For example, NASA should not pursue incremental research that is of such low risk that industry could easily complete the research. Nor should NASA pursue research of great theoretical promise if the scientific and technical hurdles are so high that it has very little chance of success.

The NASA aeronautics program includes little in the way of substantial flight tests.[2] Many research tasks focus on tool development, and not on tool validation and/or the use of validated tools to develop new system concepts. This is, in large part, an unavoidable consequence of trying to address a wide range of R&T challenges with resources that are inadequate to the task. The current approach is very broad, but it is not very deep. In addition, although NASA is conducting valuable aeronautics research, in many cases comparable work and capabilities also exist elsewhere. As a result, too much of NASA's aeronautics research has limited potential to advance the state of the art far enough and with enough urgency to make a substantial difference in meeting individual R&T challenges or the larger goal of achieving the strategic objectives of the *Decadal Survey of Civil Aeronautics*.

Furthermore, a research strategy that focuses on maintaining core competencies in areas where expertise already exists may be a reasonable approach to a tightly constrained fiscal environment. However, this approach could also trap NASA into conducting some research that is obsolete or low-priority simply because the research makes use of current expertise and capabilities. This approach also limits NASA's ability to take a leading position in new fields where it might be appropriate for the NASA of the future to possess new core competencies and new unique capabilities. For example, unmanned air vehicles (UAVs) are an emerging presence in civil aviation, and NASA should develop new capabilities related to UAV operations, safety, and control in civil airspace (see R&T challenges D10, E2, and E3).

Finding. The NASA aeronautics program has the technical ability to address each of the highest-priority R&T challenges from the *Decadal Survey of Civil Aeronautics* individually (via in-house research and/or partnerships with external research organizations). However, a substantial increase in the ARMD budget would be necessary to address all 51 challenges in a thorough and comprehensive manner. In addition, NASA's aeronautics research program faces many constraints (in terms of overall budget, the existing set of NASA centers, limitations on the ability to transfer staff positions among centers, and

[2]A notable exception concerns work related to R&T challenge A4a, Aerodynamic designs and flow-control schemes to reduce aircraft and rotor noise. NASA is planning to initiate a new phase of the Quiet Technology Demonstrator Program that involved NASA, manufacturers, and airlines in flight testing of advanced noise control techniques for subsonic fixed-wing aircraft. This is an important example of collaboration to facilitate technology transition to in-flight use.

limitations on the ability to compete with the private sector in terms of financial compensation in some critical fields), and attempting to address too many research objectives will severely limit the ability to advance the state of the art or develop new core competencies and unique capabilities that may be vital to the future of U.S. aeronautics.

Recommendation. The NASA Aeronautics Research Mission Directorate should ensure that its research program substantively advances the state of the art and makes a significant difference in a time frame of interest to users of the research results by (1) making a concerted effort to identify the potential users of ongoing research and how that research relates to those needs and (2) prioritizing potential research opportunities according to an accepted set of metrics. In addition, absent a substantial increase in funding and/or a substantial reduction in other constraints that NASA faces in conducting aeronautics research (such as facilities, workforce composition, and federal policies), NASA, in consultation with the aeronautics research community and others as appropriate, should redefine the scope and priorities within the aeronautics research program to be consistent with available resources and the priorities identified in (2) above (even if all 51 highest-priority R&T challenges from the *Decadal Survey of Civil Aeronautics* are not addressed simultaneously). This would improve the value of the research that the aeronautics program is able to perform, and it would make resources available to facilitate the development of new core competencies and unique capabilities that may be essential to the nation and to the NASA aeronautics program of the future.

GAP BETWEEN PROJECT REFERENCE DOCUMENTS AND PROJECT STRUCTURE

NASA has developed a reference document for each of its 10 aeronautics research projects. The reference documents are intended to define the rationale, scope, and detailed content of a comprehensive research effort to address each project area, but NASA does not consider them to be completed research plans.

The reference documents diagram projects in terms of a four-level hierarchy, as follows:

Level 1. Foundational physics and modeling.
Level 2. Discipline-level capabilities.
Level 3. Multidisciplinary capabilities.
Level 4. System design.

For example, the diagrams for the Subsonic Fixed Wing Project and Supersonics Project appear in Figures 4-1 and 4-2, respectively. Knowledge and capabilities are expected to flow up, from Level 1 to Level 4 (as shown), while requirements and needs are expected to flow down, from Level 4 to Level 1 (not shown). The four-level diagrams and the reference documents as a whole provide a conceptual description of the projects, but in some cases they are difficult to correlate to the manner in which the projects are being implemented. In most cases, each project is divided into research task areas, each of which is managed by an associate principal investigator (API). In some cases, there is a good correlation between the project management structure and the project level diagram. For example, the Subsonic Fixed Wing Project has 10 Level 2 areas, and an API has been appointed for each of these areas (except for power, because work in that area has been deferred), as shown in Table 4-1.

In many cases, however, it is difficult to map significant areas of research (as indicated by the API areas of responsibility) into the level diagram for the corresponding project. For example, the Supersonics Project has 7 Level 2 areas and 10 APIs (see Table 4-2). The areas of responsibility for many APIs

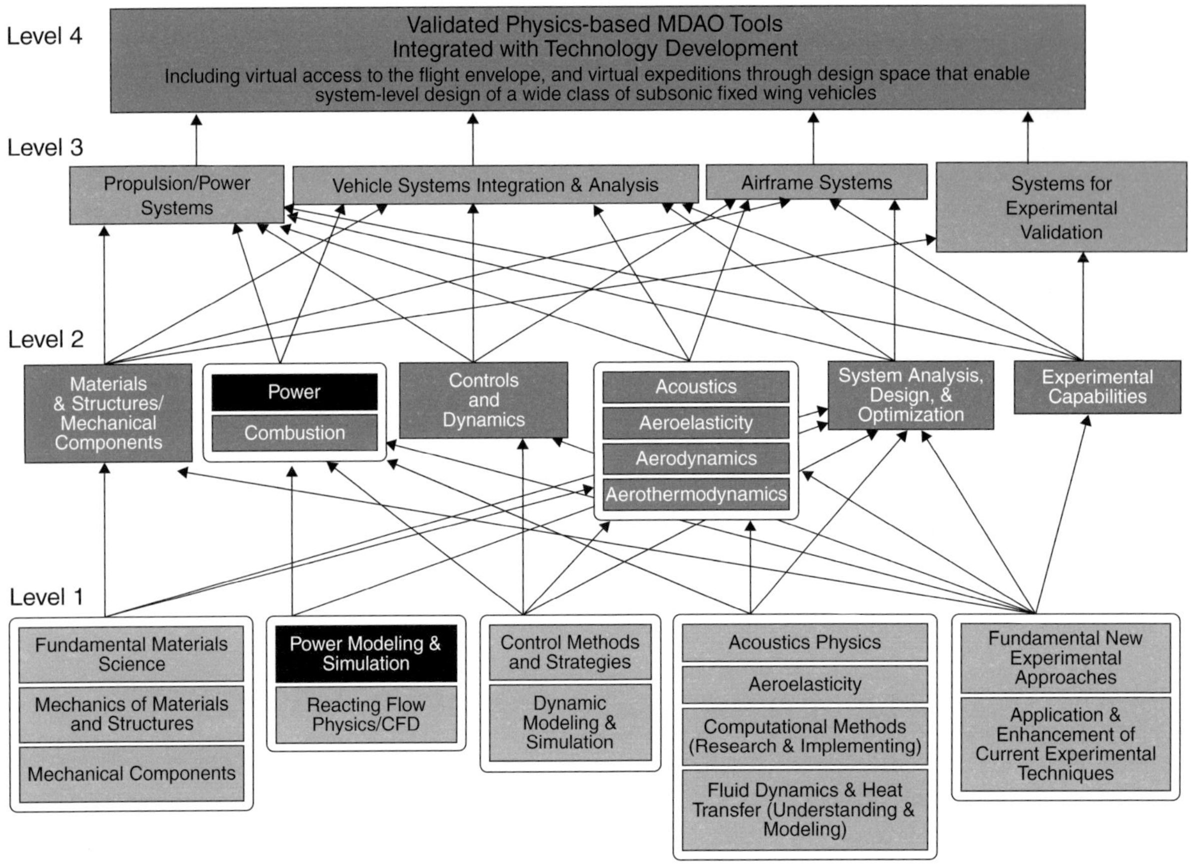

FIGURE 4-1 Subsonic Fixed Wing Project Level 1 to Level 4 integration diagram (topics shown with black background are currently deferred). Note: MDAO, Multidisciplinary Design, Analysis, and Optimization; CFD, computational fluid dynamics. SOURCE: NASA (2006a).

TABLE 4-1 Associate Principal Investigator (API) Areas of Responsibility for Level 2 Research Areas for the Subsonic Fixed Wing Project

Subsonic Fixed Wing Project: API Areas of Responsibility	Level 2 Areas
Materials and Structures	Materials and Structures/Mechanical Components
Combustion	Combustion
Controls and Dynamics	Controls and Dynamics
Acoustics	Acoustics
Aeroelasticity	Aeroelasticity
Aerodynamics	Aerodynamics
Aerothermodynamics	Aerothermodynamics
Systems Analysis, Design, and Optimization	System Analysis, Design, and Optimization
Experimental Capabilities	Experimental Capabilities
N/A	Power

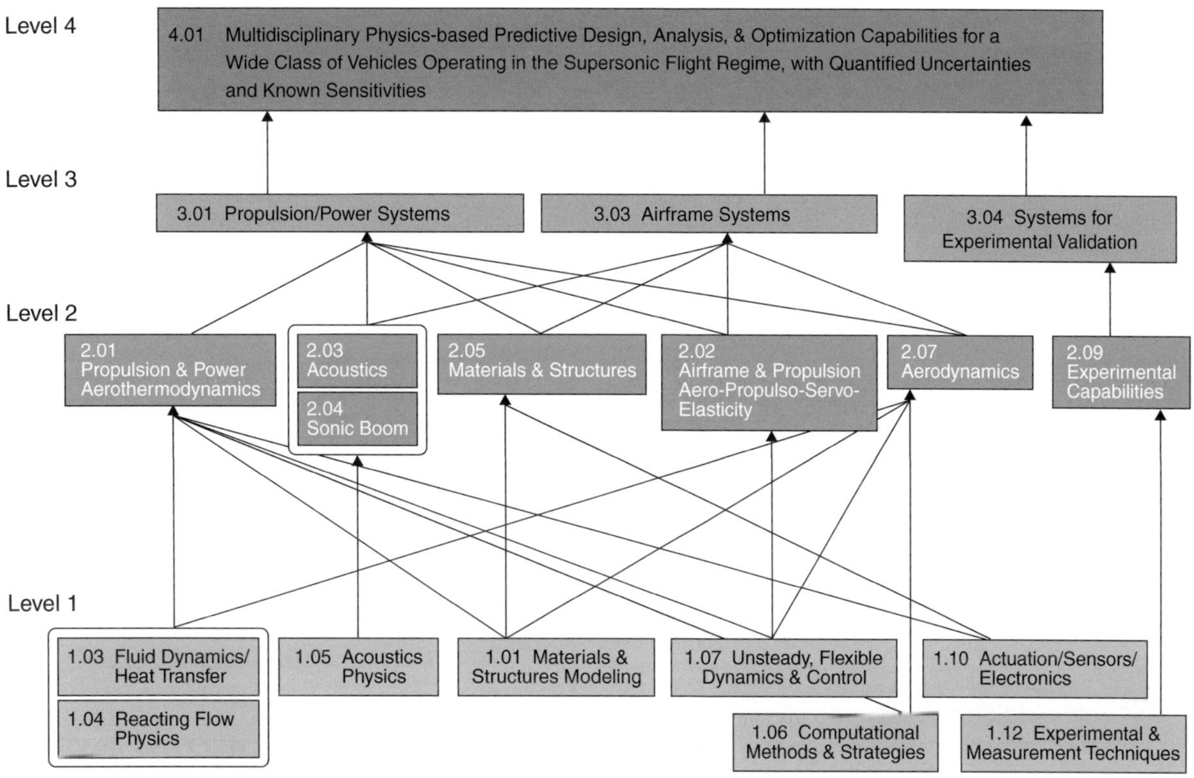

FIGURE 4-2 Supersonics Project Level 1 to Level 4 integration diagram. SOURCE: NASA (2006b).

TABLE 4-2 Associate Principal Investigator (API) Areas of Responsibility for Level 2 Research Areas for the Supersonics Project

Supersonics Project: API Areas of Responsibility	Level 2 Areas
Cruise Efficiency—Propulsion	Propulsion and Power Aerothermodynamics
Cruise Efficiency—Airframe	Aerodynamics
Airport Noise	Acoustics
Sonic Boom Modeling	Sonic Boom
Lightweight and Durability at High Temperature	Materials and Structures
Aero-propulso-servo-elasticity (APSE)	Airframe and Propulsion APSE
Experimental Validations and Capabilities	Experimental Capabilities
Systems Integration and Assessment	N/A
High-Altitude Emissions	N/A
Entry, Descent, and Landing (of spacecraft)	N/A

do not seem to cleanly map into the level diagram, and in many cases the scope of many Level 2 areas is broader than that of the closest matching API areas of responsibility.

Recommendation. As reference documents and project plans are revised and updated, NASA should continue to improve the correlation between (1) the reference documents that describe project rationale and scope and (2) the project plans and actual implementation of each project.

LOOKING FORWARD

NASA has a critical part to play in preserving the role of the United States as a leader in aeronautics. NASA research facilities and expertise support research by other federal agencies and industry, and the results of research conducted and/or sponsored by NASA are embodied in key elements of the U.S. air transportation system, military aviation, and space program. NASA aeronautics research will carry on this tradition as long as its research is properly prioritized and research tasks are executed with enough depth and vigor to produce meaningful results in a timely fashion.

REFERENCES

NASA (National Aeronautics and Space Administration). 2006a. Fundamental Aeronautics Program, Subsonic Fixed Wing Project Reference Document. Washington, D.C.: NASA Headquarters, Aeronautics Research Mission Directorate. Available online at <www.aeronautics.nasa.gov/nra_pdf/sfw_proposal_c1.pdf>.

NASA. 2006b. Fundamental Aeronautics Program, Supersonics Project Reference Document. Washington, D.C.: NASA Headquarters, Aeronautics Research Mission Directorate. Available online at <www.aeronautics.nasa.gov/nra_pdf/sup_proposal_c1.pdf>.

NRC (National Research Council). 2006. Decadal Survey of Civil Aeronautics: Foundation for the Future. Washington, D.C.: The National Academies Press. Available online at <http://www.nap.edu/catalog.php?record_id=11664>.

NSTC (National Science and Technology Council). 2006. National Aeronautics Research and Development Policy. Washington, D.C.: Office of Science and Technology Policy. Available online at <www.ostp.gov/html/NationalAeroR&DPolicy12-19-06.pdf>.

NSTC. 2007. National Plan for Aeronautics Research and Development and Related Infrastructure. Washington, D.C.: Office of Science and Technology Policy. Available online at <www.aeronautics.nasa.gov/releases/12_21_07_release.htm>.

Porter, Lisa. 2007. Aeronautics Research Mission Directorate Briefing to the National Research Council. Dr. Lisa Porter, Associate Administrator for Aeronautics. April 16.

Appendixes

Appendix A

Statement of Task

The Committee for the Assessment of NASA's Aeronautics Research Program was tasked with executing the following statement of task.

The National Research Council will assemble a committee of approximately 15 technical experts to conduct an independent assessment of NASA's fundamental aeronautics research. This assessment will include a comparison of the current NASA Aeronautics Research Mission Directorate (ARMD) research program with the future fundamental research needs identified in the recently completed *Decadal Survey of Civil Aeronautics*. Fundamental aeronautics research not addressed or highlighted by the *Decadal Survey* but considered part of the ARMD research mission will also be assessed. The scope of this assessment will include the fundamental research being conducted in the entire ARMD program portfolio, including the Fundamental Aeronautics Program, the Aviation Safety Program, and the Airspace Systems Program. For purposes of clarity, NASA uses the term "fundamental research" to include continued long-term, scientific study in areas such as physics, chemistry, materials, experimental techniques, and computational techniques that leads to a furthering of our understanding of the underlying principles that form the foundation of the core aeronautics disciplines, as well as that research that integrates the knowledge gained in these core areas to significantly enhance our capabilities, tools, and technologies at the disciplinary (e.g., aerodynamics, combustion, trajectory prediction uncertainty) and multidisciplinary (e.g., airframe design, engine design, airspace modeling and simulation) levels.

The committee will focus its assessment around the following questions:

1. How well does NASA's research portfolio implement appropriate recommendations and address relevant high-priority research and technology challenges identified in the *Decadal Survey of Civil Aeronautics*? If gaps are found, what steps should be taken by the federal government to eliminate them?

2. How well does NASA's aeronautics research portfolio address the aeronautics research requirements of NASA, particularly for robotic and human space exploration? How well does NASA's aeronautics research portfolio address other federal government department/agency non-civil aeronautics research needs? If gaps are found, what steps should be taken by NASA and/or other parts of the federal

government to eliminate them? In order to answer this question the committee will identify and prioritize requirements for such research that fall within the scope of NASA's Aeronautics Research Program. To assist in the identification of such research requirements, NASA will provide the NRC with a list of its current research activities that contribute to these areas no later than March 12, 2007. It is likely that much of this research will be "dual use" or even "triple use," meaning that the research may provide benefit to the civil aeronautics community, and/or the space exploration community, and/or departments and agencies with non-civil aeronautics research needs.

3. Will the nation have a skilled research workforce and research facilities commensurate with the requirements in (1) and (2) above? What critical improvements in workforce expertise and research facilities, if any, should NASA and the nation make to achieve the goals of NASA's research program?

Appendix B

Biographies of Committee Members

CARL J. MEADE, *Co-chair,* is director of the Lunar Lander Program and director of Space Systems at Northrop Grumman Integrated Systems. He and his team are responsible for both government and civil projects relating to crewed spaceflight and nonpayload military space vehicles. He was previously employed at Lockheed Martin Aeronautics Company (a.k.a. the Skunk Works) in Palmdale, California, where he directed the development of various advanced aerospace vehicles, including the X-33. Mr. Meade is a former Air Force fighter and experimental test pilot and NASA astronaut, having complete three space shuttle flights. He has particular expertise in risk/benefit analysis and decision making, modeling and simulation, signal processing, flying qualities, human factors, and avionics.

DONALD W. RICHARDSON, *Co-chair,* retired in 2005 as the vice president of Science Applications International Corporation responsible for all Federal Aviation Administration (FAA) and civil aviation corporate activities. He is a fellow and past president of the American Institute of Aeronautics and Astronautics (AIAA), of which he has been a member for 58 years. He is also a fellow of the Royal Aeronautical Society and served on its Engineering Council. He was awarded the NASA Public Service Medal in 2002 for his work in reinvigorating U.S. federal funding for aeronautics research and development (R&D). He holds bachelor's, master's and Ph.D. degrees in aeronautical and mechanical engineering. A commercial instrument pilot with multiengine land and seaplane ratings, Dr. Richardson has been an active pilot for 59 years. His engineering career has included assignments as an aerodynamics and flight-test engineer, research pilot, and engineering manager.

RICHARD ABBOTT is a technical fellow emeritus at Lockheed Martin Aeronautics Company in Palmdale, California. He received a Ph.D. in chemical physics from Northern Illinois University, where his research concentrated on cooperative phenomena in molecular systems. He continued studies as a research associate in statistical mechanics at the University of Chicago's James Franck Institute, where he contributed to theories of energy relaxation in condensed media using Monte Carlo and molecular dynamics techniques. Dr. Abbott's aerospace career includes more than 25 years of experience in the areas of guidance, navigation, and control systems design and analysis; sensor data fusion design; and sensor system simulation and modeling for both manned and unmanned aircraft. He has supervised the development and execution of large-scale simulations of complex air vehicles, led the development of the avionics functional architecture for the demonstration and validation phase of the YF-22 program, and developed fault-detection and redundancy-management algorithms for navigation systems aboard the X-33 single-stage-to-orbit vehicle. He also has served as principal investigator for the Defense Advanced Research Projects Agency (DARPA) software-enabled control technologies for reliable autonomous control project and has been the co-chair for the Technologies for Autonomous Control session of the IEEE Aerospace Conferences.

MEYER J. (MIKE) BENZAKEIN (NAE) is the chair of the Aerospace Engineering Department at Ohio State University. Dr. Benzakein began his professional career at General Electric in 1967, where he subsequently served in a number of positions in advanced technology and project and product engineering. He led the CFM56 Engineering Program from 1984 to 1993 and the GE90 Engineering Program from 1993 to February 1995. Dr. Benzakein in February 1995 became general manager for engine systems design and integration, and in this capacity he had the responsibility for engineering leadership and technical oversight of GE Evendale Commercial and Military Aircraft Engines. In January 1996, Dr. Benzakein took over the position of general manager, advanced engineering. He was responsible for leading technology development and certification/qualification of new engine products. His charter is to ensure that the customer expectations as well as the needs of GE Aircraft Engines (GEAE) Multi-generation Product Plans are met. Dr. Benzakein is responsible for GEAE front-end initiatives in driving technology maturation, strengthening the linkage between preliminary design, engine systems design, and production hardware design. He was elected a member of the National Academy of Engineering in 2001. That year he received the Gold Medal Award from the Royal Aeronautical Society. He is a fellow of the Royal Aeronautical Society and the American Society of Mechanical Engineers, and he is the 2007 recipient of the AIAA's Reed Aeronautics Award, which is the highest honor that the AIAA bestows for achievements in aeronautical science and engineering.

JOHN T. (TOM) BEST is the director of the Capabilities Integration Directorate at the Air Force's Arnold Engineering Development Center (AEDC), which contains the largest complex of ground aerospace test facilities in the world. AEDC provides developmental testing of propulsion, aircraft, missile, and space systems for the U.S. government, industry, and foreign governments. Mr. Best is responsible for capabilities assessment and planning, technology development and transfer, and intelligence integration. Mr. Best also acts as the AEDC leader for NASA/Department of Defense (DoD) collaboration under the National Partnership for Aeronautical Testing (NPAT). Mr. Best served in the past as head of the Long Range Requirements and Facility Planning Branch, the chief of the Applied Technology Division, and deputy director of the 704th Test Group. He also served for one-and-a-half years as a staff specialist in the Office of the Deputy Director for Defense Research and Engineering (Test and Evaluation) in Washington D.C., overseeing the work of the DoD's Major Range and Test Facility Base. Currently, he also serves as the technical project officer for a data exchange agreement on wind tunnels with Germany.

IAIN D. BOYD is a professor of aerospace engineering at the University of Michigan. He leads a research group that develops and applies physical models and numerical methods for the simulation of nonequilibrium gas flows and plasmas. Current application areas include electric propulsion for spacecraft, hypersonic aerothermodynamics, and flows involving microelectromechanical systems, and deposition of thin films for advanced materials. After receiving a Ph.D. in aeronautics and astronautics from the University of Southampton, England, Dr. Boyd was a research scientist at NASA Ames

Research Center for 4 years. He was subsequently an associate professor of mechanical and aerospace engineering at Cornell University, before joining the faculty of the University of Michigan in 1999. Dr. Boyd received the AIAA Lawrence Sperry Award in 1998, and he is an associate editor of the AIAA *Journal of Spacecraft and Rockets*. He has authored more than 300 research papers.

AMY L. BUHRIG is director of technology for Boeing Commercial Airplanes. She is responsible for leading the definition of technology required to enable future products and services, while ensuring that the company's investments are aligned with business unit strategy and industry economics. Integral to these responsibilities is working with organizations to develop the skills, processes, and tools necessary to enable the application of future technologies. Ms. Buhrig is also the primary interface between Boeing Commercial Airplanes and Phantom Works, the company's R&D organization, to maximize the value derived from the company's R&D activities. Ms. Buhrig has also worked at the Phantom Works, most recently leading a team to define the strategy for the Structural Technologies, Prototyping, and Quality organization. Other Phantom Works assignments included understanding enterprise-wide technology needs in order to leverage venture capital investments, and pursuing R&D contracts for the Mathematics and Computing Technology organization. The first 20 years of her career were spent in Boeing Integrated Defense Systems. She performed studies to quantify the benefit of investing in novel design methods for the Boeing 777 and F-22 aircraft and assessed company strengths applicable to the commercial space market, and she was vice president of marketing and sales for Boeing's Sea Launch Company.

DAVID E. (ED) CROW (NAE) joined the Department of Mechanical Engineering at the University of Connecticut as a distinguished professor-in residence in 2002, following a long career with Pratt & Whitney, where he rose to the position of senior vice president for the Engineering organization. As such, he was responsible for the design, development, validation, and certification of all Pratt & Whitney large commercial engines, military engines, and rocket products. He also led the research and development of advanced technologies systems to meet future aircraft requirements. Dr. Crow also served as senior vice president for the Large Commercial Engines organization, which included the PW4000 and JT9D high-thrust family of products. He was elected a member of the National Academy of Engineering for his leadership in the engineering design of high-bypass-ratio gas turbine aircraft engines.

FRANK L. FRISBIE is vice president of strategic planning at Apptis, Inc. He was a longtime former senior executive with the Department of Defense and the Federal Aviation Administration and more recently vice president and senior client executive for civil aviation with Northrop Grumman Information Technology. He joined the FAA in 1958, where he held a variety of positions. In his last two FAA posts, he was directly responsible for research, development, system engineering, acquisition, deployment, and maintenance of all 20,000 air traffic control facilities in the United States. Mr. Frisbie was awarded the prestigious Glen A. Gilbert Memorial Award in 2002 by the Air Traffic Control Association for his longstanding contributions to the air traffic control and civil aviation communities. He has been involved with the development, deployment, maintenance, and operation of virtually every system employed in the U.S. civil aviation infrastructure. He is currently a member of the board of governors for the Center for Unmanned Air Systems Integration in the National Airspace System. He earned his B.E.E. degree from Manhattan College and his M.B.A. degree from American University. He is a member of the Russian Academy of Navigation and Motion Control, holds a professional engineer's license, and is a frequent contributor to the *Journal of Air Traffic Control,* for which he writes on contemporary air traffic management technical and policy issues.

EPHRAHIM GARCIA is an associate professor of mechanical and aerospace engineering at Cornell University, where his interests lie in the development of new types of actuation systems utilizing smart-material transducers, system-level demonstrations of smart structures applied to defense platforms, morphing aircraft systems, and bio-inspired intelligent machines. Dr. Garcia served as a program manager in the Defense Sciences Office at DARPA from 1998 to 2002. From 1991 to 1998, he was an assistant and associate professor of mechanical engineering at Vanderbilt University, where he was director of the Center for Intelligent Mechatronics and the Smart Structures Laboratory. In this capacity he directed research in the areas of smart structures, control structure interaction, and bioinspired robotics. From 1991 to 1997, he owned and operated Garman Systems, Inc. (now Dynamic Structures and Materials, LLC), a small engineering corporation that designed and fabricated devices in adaptive structural systems utilizing piezoelectric, electrostrictive, and shape memory alloy materials. Dr. Garcia was named an Office of Naval Research Young Investigator, appointed a 1993 Presidential Faculty Fellow by President Clinton, and received Summer Faculty Fellowship awards from the Air Force on four occasions. In 1995, he was named Most Promising Scientist by *Hispanic Engineer* magazine. In 2002, Dr. Garcia received the American Society of Mechanical Engineers' (ASME's) Adaptive Structures Prize for "significant contributions to the sciences and technologies associated with adaptive structures and/or materials systems." Dr. Garcia is author of more than 175 articles, book chapters, edited volumes, and a textbook entitled *Mechanics of Microelectromechanical Systems*. He is the chair of the ASME Aerospace Division's Executive Committee and was recently appointed editor-in-chief of *Smart Materials and Structures*.

PRABHAT HAJELA is a professor of mechanical, aerospace, and nuclear engineering at Rensselaer Polytechnic Institute (RPI). His current research interests include the analysis and design optimization of structural and multidisciplinary systems, system reliability, emergent computing paradigms for design, artificial intelligence, and machine learning in multidisciplinary analysis and design. Dr. Hajela recently completed a year as an ASME congressional fellow in the office of Senator Conrad Burns, advising on technology policy. Before joining RPI, he was on the faculty at the University of Florida for 7 years. Dr. Hajela is a fellow of the AIAA, the Aeronautical Society of India, and the ASME, and he is vice president of the International Society of Structural and Multidisciplinary Optimization. He has served on the Multidisciplinary Design Optimization Technical Committee of the AIAA and the executive committee for the ASME Aerospace Division (chair, 2001-2002) and was chair of the Division's Technical Committee on Structures and Materials (1999-2002). He has served as editor of *Evolutionary Optimization* and as associate editor of the *AIAA Journal,* and he is on the editorial board of six other international journals. He has published more than 260 papers and articles in the areas of structural and multidisciplinary optimization and is an author or coauthor of four books in these areas. Dr. Hajela has an M.S. in mechanical engineering and a Ph.D. in aeronautics and astronautics from Stanford University and a B.Tech. in aeronautical engineering from the Indian Institute of Technology in Kanpur, India.

JOHN B. HAYHURST retired in 2004 as senior vice president of The Boeing Company and president of Boeing Air Traffic Management after 33 years at Boeing and 3½ years in these positions. Previously, Mr. Hayhurst was vice president of business development for the Commercial Airplane Services business unit of Boeing Commercial Airplanes Group (BCAG). Prior to this assignment, he served as vice president and general manager of 737 programs. In addition, he was general manager of the BCAG production site in Renton, Washington. Before that, he served as vice president for the Americas and was responsible for the Boeing business relationships with airline customers in North America and Latin America and for the sale of Boeing commercial airplanes to customers in those regions. Mr. Hayhurst joined Boeing in 1969 as a customer support engineer. He held positions of increasing responsibility related to commercial airplanes and in 1987 was promoted to vice president of marketing. In this position, he played a significant role in the launch of the Boeing 777. Subsequently, he was responsible for leading teams planning the design, development, and manufacture of aircraft larger than the Boeing 747. He then served as vice president-general manager of the Boeing 747-500X/600X program. Mr. Hayhurst is a fellow of the Royal Aeronautical Society and holds a bachelor's degree in aeronautical engineering from Purdue University. He received a master's degree in business administration from the University of Washington in 1971. In 1998, Mr. Hayhurst was awarded an honorary doctorate in engineering by Purdue University.

NANCY G. LEVESON (NAE) is professor of aeronautics and astronautics and professor of engineering systems at the Massachusetts Institute of Technology. Dr. Leveson conducts research on system safety, software engineering, system engineering, and human–computer interaction. In 1999, she received the Association for Computing Machinery's (ACM's) Allen Newell Award for outstanding computer science research and in 1995 the AIAA Information Systems Award for "developing the field of software safety and for promoting responsible software and system engineering practices where life and property are at stake." In 2004 Dr. Leveson received the ACM Sigsoft Outstanding Research Award. She has published more than 200 research papers and is author of a book, *Safeware: System Safety and Computers* (Addison-Wesley). She has served on numerous National Research Council committees.[1]

ELI RESHOTKO (NAE) graduated from the California Institute of Technology with a Ph.D. in aeronautics and physics. Dr. Reshotko is currently the Kent H. Smith Professor Emeritus of Engineering at Case Western Reserve University. He was elected to the National Academy of Engineering in 1984 and is a fellow of the following societies: AIAA, ASME, the American Physical Society, and the American Academy of Mechanics, which he served as president. He is coauthor of more than 100 publications and is affiliated with many task forces, committees, and governing boards, on several of which he served as chair. His area of expertise is viscous effects in external and internal aerodynamics; two- and three-dimensional compressible boundary layers and heat transfer; stability and transition of viscous flows, both incompressible and compressible; and low-drag technology for aircraft and underwater vehicles. He has expertise in propulsion engineering, thermodynamics, aerodynamics, and aircraft propulsion.

RAYMOND (RAY) VALEIKA retired from Delta as senior vice president–technical operations (TechOps). He directed a worldwide maintenance and engineering staff of more than 10,000 professionals, maintaining a fleet of nearly 600 aircraft. Currently, he is an independent consultant advising major companies on aviation matters and an internationally recognized senior airline operations executive with more than 40 years of managing the maintenance operations of large airlines. Through Mr. Valeika's leadership and focus on the continuous improvement of the human processes in aviation maintenance, Delta TechOps consistently rated at the top of the industry for performance benchmarks in the areas of safety, quality, productivity, and reliability. Mr. Valeika was honored with the Air Transport Association's Nuts and Bolts award, recognizing his leadership in the aviation industry. Finally, his leadership of the human side has been recognized over the years with a Humanitarian Award from the Community Mayors of New York, New Jersey, and Connecticut, and a Laurel award from *Aviation Week and Space Technology* for his role with human factors training at Continental. In October 1999, Mr. Valeika received the Marvin Whitlock Award from the Society of Automotive Engineers. Most recently, the Aviation Week Group

[1]Dr. Leveson resigned from the committee in May 2007.

honored him with a lifetime achievement award. He is currently a member of the National Research Council's Aeronautics and Space Engineering Board. Previously, he held senior executive positions with Pan Am and Continental Airlines as well as Delta. He graduated from St. Louis University with a degree in aeronautical engineering in 1964.

Appendix C

Validating the Ranking of the Research and Technology Challenges from the *Decadal Survey*

This appendix investigates whether the selection of the 51 highest-priority research and technology (R&T) challenges in the *Decadal Survey of Civil Aeronautics* (NRC, 2006) remains valid in light of the *National Aeronautics Research and Development Policy* (NSTC, 2006) and the *National Plan for Aeronautics Research and Development and Related Infrastructure* (NSTC, 2007).

The *Decadal Survey* ranks 89 R&T challenges according to their ability to accomplish strategic objectives for U.S. aeronautics research. At the time that study was conducted, the federal government had yet to define what those strategic objectives should be. Therefore, in order to conduct the ranking, the steering committee that oversaw the *Decadal Survey* identified and defined six strategic objectives that, in its estimation, should motivate and guide the next decade of civil aeronautics research in the United States, pending the release of a national research and development (R&D) plan for aeronautics.

After the National Research Council published the *Decadal Survey of Civil Aeronautics,* the National Science and Technology Council released the *National Aeronautics Research and Development Policy* and the more detailed National Plan. As shown in Table C-1, there is excellent correlation between the strategic objectives of the *Decadal Survey* and the key principles of the National Policy and the National Plan. The only exceptions are "support for the space program" (which appears in the *Decadal Survey*) and "world-class aeronautics workforce" (which appears in the National Policy and the National Plan, although the National Plan includes no research goals or objectives related to this principle).[1]

To assess how the list of the top 51 R&T challenges might have looked if the National Policy and the National Plan had predated the *Decadal Survey of Civil Aeronautics,* the committee for the present study considered the following factors:

1. In the process of ranking the R&T challenges, the *Decadal Survey of Civil Aeronautics* uses different weightings for different strategic objectives. A weighting of 5 was given to capacity and to

[1]The *National Plan for Aeronautics Research and Development and Related Infrastructure* notes that "aerospace workforce issues are being explored by the Aerospace Revitalization Task Force led by the Department of Labor pursuant to Public Law 109-420" (NSTC, 2007, p. 2).

TABLE C-1 Comparison of the Strategic Objectives from the *Decadal Survey of Civil Aeronautics* with the Principles from the *National Aeronautics Research and Development Policy* and the *National Plan for Aeronautics Research and Development and Related Infrastructure*

Strategic Objectives: *Decadal Survey*[a]	Principles: National Policy[b] and National Plan[c]
• Increase capacity.	• Mobility through the air is vital to economic stability, growth, and security as a nation.
• Improve safety and reliability.	• Aviation safety is paramount.
• Increase efficiency and performance.	• Assuring energy availability and efficiency is central to the growth of the aeronautics enterprise.
• Reduce energy consumption and environmental impact.	• The environment must be protected while sustaining growth in air transportation.
• Take advantage of synergies with national and homeland security.	• Aviation is vital to national security and homeland defense. • Security of and within the aeronautics enterprise must be maintained.
• Support the space program.	
	• The United States should continue to possess, rely on, and develop its world-class aeronautics workforce.

[a]NRC (2006), p. 1.
[b]NSTC (2006), pp. 7-8.
[c]NSTC (2007), pp. 1-2.

safety and reliability, 3 was given to efficiency and performance and to energy consumption and environmental impact, and 1 was given to synergies with national and homeland security and to support for the space program. The National Policy and the National Plan, however, treat all the principles as equally important.

2. In the National Plan, the principles concerning (a) energy availability and efficiency and (b) the environment have been combined.
3. Support to the space program is not included as a principle of the National Policy, and the National Plan includes no R&D goals or objectives related to space except with regard to the demonstration of sustained, controlled hypersonic flight in support of national defense and homeland security (including space launch applications).
4. NASA is directed elsewhere in the National Policy to maintain its core competencies, including core competencies "that support NASA's human and robotic space activities" (NSTC, 2006, p. 9).

Accordingly, the committee for this study reranked the R&T challenges of the *Decadal Survey* by changing the weightings as follows:

1. As a result of factor 1, above, all strategic objectives were assigned the same value (1.0), except that:
2. As a result of factor 2, the strategic objectives corresponding to (1) energy availability and efficiency and (2) the environment were combined by giving them combined weighting of 1.0 (i.e., a weighting of 0.5 each), and

3. As a result of factors 3 and 4, the strategic objective for support to space was also given a smaller weighting (0.5). Factor 3 implied the value should be smaller than the value assigned to the other strategic objectives, and factor 4 implied the value should be greater than zero.

In other words, the priorities of the 89 R&T challenges from the *Decadal Survey of Civil Aeronautics* were recalculated using the following weightings:

Weighting	*Strategic Objective*
1.0	Increase capacity
1.0	Improve safety and reliability
0.5	Increase efficiency and performance
0.5	Reduce energy consumption and environmental impact
1.0	Take advantage of synergies with national and homeland security
0.5	Support the space program

Using this different weighting scheme, only four R&T challenges not previously ranked among the top 51 moved into the top 51. Those challenges are as follows:

- A12. Accurate predictions of thermal balance and techniques for the reduction of heat transfer to hypersonic vehicles
- A14. Efficient control authority of advanced configurations to permit robust operations at hypersonic speeds and for access-to-space vehicles
- B12. Hypersonic hydrocarbon-fueled scramjet
- E12. Autonomous flight monitoring of manned and unmanned aircraft[2]

The committee for this study does not assert that these four R&T challenges should be considered among NASA's highest-priority challenges, in part because the ranking of challenges by the *Decadal Survey* was much more than a mathematical exercise. Rather, the committee concludes that this exercise confirms the validity of rankings in the *Decadal Survey,* because it seems clear that even if the National Policy and the National Plan had been issued prior to the *Decadal Survey,* the rankings in the *Decadal Survey* would be essentially the same, with perhaps just a few R&T challenges near the cut-off point changing sides.

This study determined how effectively NASA is addressing the four R&T challenges identified above. This information is provided with the caveat that shortcomings in addressing these challenges are not necessarily shortcomings in NASA's aeronautics research program, given that, unlike the challenges discussed in Chapter 2, these challenges do not appear among the highest-priority R&T challenges in the *Decadal Survey of Civil Aeronautics.*

The assessment of these four challenges uses the same color-coded grading scheme that is described in detail at the beginning of Chapter 2. Briefly stated, a green grade means that the project will substantively advance the state of the art with no significant shortcomings. A yellow grade means that the

[2]This reranking would displace the following R&T challenges out of the top 51:

- A7b. Accuracy of wake vortex prediction, and vortex detection and mitigation techniques
- A11. Robust and efficient multidisciplinary design tools
- B9. High-reliability, high-performance, and high-power-density aircraft electric power systems
- E7. Adaptive ATM techniques to minimize the impact of weather by taking better advantage of improved probabilistic forecasts

project contains minor shortcomings, which are recoverable within the current overall project concept. A black grade means that the project contains major shortcomings, which would be difficult to recover from within the current overall project concept. White means that the challenge is not relevant.

A12 Accurate predictions of thermal balance and techniques for the reduction of heat transfer to hypersonic vehicles

SFW	SRW	Supersonics	Hypersonics	Airportal	Airspace	IVHM	IIFD	IRAC	Aging A/C

This R&T challenge has the following milestones:[3]

- Improve models for predicting the effects of ablation on heat transfer.
- Develop a high-fidelity model for radiating shock layers.
- Develop turbulence models validated against experimental data for highly cooled walls.

The Aerodynamics, Aerothermodynamics, and Plasmadynamics element of the Hypersonics Project includes tasks for developing high-fidelity thermal radiation models. The same element also includes a small activity on ablation modeling. The Hypersonics Project does not address turbulence modeling for highly cooled walls.

A14 Efficient control authority of advanced configurations to permit robust operations at hypersonic speeds and for access-to-space vehicles

SFW	SRW	Supersonics	Hypersonics	Airportal	Airspace	IVHM	IIFD	IRAC	Aging A/C
			Black						

This R&T challenge has the following milestones:

- Develop techniques to accurately predict flow-control authority in shock-dominated flows and in transitional flows.
- Demonstrate novel flow-control techniques applicable to hypersonic vehicles.
- Develop novel ground- and flight-test instrumentation techniques for validation of analytical and computational models.

The Hypersonics Project includes little substantive work related to the above milestones. Only one task in the Aerodynamics, Aerothermodynamics, and Plasmadynamics element mentions flow-control demonstration—and it is mentioned as just one of several milestones associated with that task.

[3]Milestones for each R&T challenge are defined in Appendixes A through E of the *Decadal Survey of Civil Aeronautics* (NRC, 2006).

B12 Hypersonic hydrocarbon-fueled scramjet

SFW	SRW	Supersonics	Hypersonics	Airportal	Airspace	IVHM	IIFD	IRAC	Aging A/C

This R&T challenge has the following milestones:

- Develop advanced instrumentation capable of measuring time-averaged and time-resolved flow parameters to validate design tools.
- Complete unit experiments based on generic inlets, isolators, combustors, and nozzles to provide to provide benchmark data sets for model validation.
- Conduct experiments on mode transition for validation of unsteady models.
- Assist DoD flight demonstration programs that are currently in progress.

The Propulsion and Experimental Capabilities elements of the Hypersonics Project include several tasks devoted to supporting the DoD's X-51 Program, which is developing and testing hydrocarbon scramjet technology. The Hypersonics Project also includes diagnostics research related to this challenge that is being conducted alongside the X-51 tests.

E12 Autonomous flight monitoring of manned and unmanned aircraft

SFW	SRW	Supersonics	Hypersonics	Airportal	Airspace	IVHM	IIFD	IRAC	Aging A/C

This R&T challenge has the following milestones:

- Produce a detailed set of requirements and design specification for flight monitoring systems deployed on manned and unmanned aircraft.
- Demonstrate algorithms and knowledge to enable a flight monitoring system that accurately anticipates, detects, and diagnoses flight plan deviations.
- Demonstrate the ability to more accurately project the near-term results of manipulating aircraft controls and inform pilots of likely consequences in terms of aircraft motion, potential collisions, airspace violations, etc.
- Design protocols for disseminating information from flight monitoring systems locally and throughout the air transportation system.
- Specify corrective actions appropriate for manned and unmanned aircraft in response to unplanned deviations detected by a flight monitoring system.

The Integrated Intelligent Flight Deck Project is supporting substantial research related to manned aircraft, but it is not dealing with unmanned aircraft.

REFERENCES

NRC (National Research Council). 2006. Decadal Survey of Civil Aeronautics: Foundation for the Future. Washington, D.C.: The National Academies Press. Available online at <http://www.nap.edu/catalog.php?record_id=11664>.

NSTC (National Science and Technology Council). 2006. National Aeronautics Research and Development Policy. Washington, D.C.: Office of Science and Technology Policy. Available online at <www.ostp.gov/html/NationalAeroR&DPolicy12-19-06.pdf>.

NSTC. 2007. National Plan for Aeronautics Research and Development and Related Infrastructure. Washington, D.C.: Office of Science and Technology Policy. Available online at <www.aeronautics.nasa.gov/releases/12_21_07_release.htm>.

Appendix D

Acronyms

3-D	three-dimensional
4-D	four-dimensional
API	associate principal investigator
APSE	aero-propulso-servo-elasticity
ARMD	Aeronautics Research Mission Directorate
ATM	air traffic management
ATP	Aeronautics Test Program
BWB	blended wing body
CFD	computational fluid dynamics
CP	challenge problem
DARPA	Defense Advanced Research Projects Agency
DoD	Department of Defense
ESTOL	extremely short takeoff and landing
FAA	Federal Aviation Administration
HMMES	High Mass Mars Entry Systems
HRRLS	Highly Reliable Reusable Launch Systems
IDA	Institute for Defense Analyses
IIFD	Integrated Intelligent Flight Deck
IRAC	Integrated Resilient Aircraft Control

IVHM	Integrated Vehicle Health Management
JPDO	Joint Planning and Development Office
MDAO	Multidisciplinary Design, Analysis, and Optimization
NASA	National Aeronautics and Space Administration
NextGen	Next Generation Air Transportation System (formerly abbreviated NGATS)
NGATS	Next Generation Air Transportation System (now commonly abbreviated as NextGen)
NO_x	oxides of nitrogen
NPAT	National Partnership for Aeronautical Testing
NRA	NASA Research Announcement
NRC	National Research Council
NSTC	National Science and Technology Council
PI	principal investigator
QFD	quality function deployment
R&D	research and development
R&T	research and technology
SCAP	Shared Capabilities Assets Program
SFW	Subsonic Fixed Wing
SRW	Subsonic Rotary Wing
T&E	testing and evaluation
UAV	unmanned air vehicle
V/STOL	vertical and short takeoff and landing